LEARNING DISABILITIES

LEARNING DISABILITIES
Contemporary Viewpoints

Edited by

Bryant J. Cratty

Professor Emeritus
University of California at Los Angeles

and

Richard L. Goldman

Learning Disabilities Consultant
Calabasas Hills, California

Foreword by Drake D. Duane, MD

Routledge
Taylor & Francis Group
LONDON AND NEW YORK

First published 1996 by Harwood Academic Publishers.

2 Park Square, Milton Park, Abingdon, Oxon OX14 4RN
711 Third Avenue, New York, NY 10017, USA

Routledge is an imprint of the Taylor & Francis Group, an informa business

First issued in hardback 2016

Transferred to Digital Printing 2004

British Library Cataloguing in Publication Data

Learning disabilities : contemporary viewpoints
1.Learning disabilities 2.Learning disabled
I.Cratty, Bryant J. II.Goldman, Richard L.
616.8'5889

ISBN 978-3-7186-0623-8 (pbk)
ISBN 978-1-138-18122-9 (hbk)

This book is dedicated to those children with learning disabilities and their families. Their persistence and courage on a daily basis in dealing with adversity is a lesson for all of us. We also admire all those professionals who devote their lives to the learning disabled and whose sensitivity, understanding, and awareness help these youngsters meet their potential.

CONTENTS

FOREWORD

No group of neurologically based disorders more urgently calls for an interdisciplinary offensive than learning disabilities. These constitutionally determined conditions sculpt a child's perception and interaction with his or her social environment, commonly before formal school years even commence. Perplexed parents and preschool educators who seek counsel from professionals in pediatrics, speech and language, or occupational therapy about specific student difficulties are only sometimes provided with sufficient insight into the behavioral disorder.

During the school years, academic demands upon the unprepared mind—in an arena filled with peers of similar age but varying levels of cognitive development—further confuse the child, frustrating optimal achievement. Too often, only after lifelong emotional wounds have been inflicted, is the nature of the student's struggle correctly identified but still under-rehabilitated educationally, emotionally, and physically. To avert this failure of an appropriate societal response, the private educational sector has stepped forward to construct an environment conducive to the academic and social growth of individual learners, as this text initiated by a private school confirms. This series of lectures, preserved in print, provides a permanent resource for the many professionals—teachers, physicians, psychologists, social workers, speech and language clinicians, and occupational therapists—who interact with learning disabled students, as well as for the students themselves and their families.

These chapters have been written by astute clinicians who appreciate the rigor of clinical research, yet are sensitive to the needs of these students. I commend the editors of this book for their foresight in creating the lecture series, the authors for their perceptiveness, and I recommend their efforts to a diverse readership as an intellectual resource and inspiration.

Drake D. Duane, MD

Director, *Institute for Developmental Behavioral Neurology/Biological Psychiatry*
Professor, *Speech and Hearing Sciences, Arizona State University*

President, *International Academy for Research in Learning Disabilities*
Chair, *Scientific Advisory Board, National Dyslexia Research Foundation*

PREFACE

This book reflects a commitment by the editors to increase public awareness of learning disabilities. Its primary goals are advancement of the frontiers of knowledge and dissemination of contemporary information to parents and professionals. The lectures contained here were given over a three-year period as part of a seminar series at Landmark West School in Encino, California.

These lectures present cutting-edge information to parents, teachers, university students, and other professionals who deal with learning disabled children on a regular basis. Original content has been edited so that others might profit, as leaders in the field interpret current research and provide contemporary clinical insights. Presentations as well as question and answer sessions are included.

Speakers were selected based on reputation and scholarship in the field. The lectures represent a broad spectrum of expertise from many disciplines: Drs. Larry Silver and Dennis Cantwell are psychiatrists recognized for their medical observations about the effects of medication on attention deficit disorders; Steven Forness is a well-known educator who deals with the various emotional aspects of learning disabilities; Bryant Cratty, Professor Emeritus, UCLA, summarizes recent research about poor motor coordination of these children; clinically oriented psychologists James Gardner and Martin Goodman explore family dynamics and the effects of school failure. University professors and researchers Frank Manis and Michael Spagna present a clear picture of current research and differences in dyslexic children that are currently emerging in the literature. The professors, psychologists, doctors, and psychiatrists share their expertise and views on family, educational, and social issues relating to children with learning disabilities.

The book concludes with a final lecture by Joan Esposito, an adult with dyslexia and attention deficit disorder, who could barely read until she was in her forties. Ms. Esposito is past president of the California Learning Disabilities Association, and executive director and founder of the Dyslexia Awareness and Resource Center in Santa Barbara, California. She has turned her personal experience into a crusade, at the state and national level, by becoming a passionate advocate for dyslexic children and adults. Through her descriptions, we can all better understand the frustrations and challenges of having a learning disability.

The editors hope that this book will not only alert readers to current ideas and trends in the field but, more importantly, create a better understanding and awareness of learning disabilities and attention deficit disorders.

ACKNOWLEDGMENTS

Several individuals provided helpful assistance in the writing of this book, and we would like to express our thanks. Debra McCowen provided expert editorial help in obtaining cogent essays from videotaped speeches. Mark Oechsli assisted in the videotaping of many speakers. Maria Farnsworth's skills were invaluable during the final drafting of the chapters. In addition, the editors appreciate the work done by Blessy Briscoe in helping to design the cover of this publication.

CONTRIBUTORS

Dennis P. Cantwell, MD University of California, Los Angeles, Joseph Campbell Professor of Child Psychiatry; Director of Residency in Child Psychiatry

Bryant J. Cratty, EdD University of California, Los Angeles, Professor Emeritus of Kinesiology

Joan T. Esposito Dyslexia Awareness and Resource Center, Founder and Executive Director; California Learning Disabilities Association, Past President

Steven Forness, EdD University of California, Los Angeles, Professor of Psychiatry and Biobehavioral Sciences, School of Medicine; Principal of In-Patient School, Neuropsychiatric Institute

James Gardner, PhD Clinical Psychologist

A. Martin Goodman, PhD Clinical Psychologist

Franklin R. Manis, PhD University of Southern California, Professor of Psychology

Larry B. Silver, MD Georgetown University School of Medicine, Clinical Professor of Psychiatry; Director of Training in Child and Adolescent Psychiatry

Michael E. Spagna, PhD California State University, Northridge, Assistant Professor of Special Education

The following speech was given by Joanna Sadowsky, a student graduating from Landmark West School, Encino, California.

GRADUATION SPEECH

This speech is dedicated to anyone who ever has or ever will experience the difficulties of a learning disability.

It's amazing to me when I remember how I used to be, how I felt and how things were. How the struggle tugged at my heart making me scared and vulnerable. How most often I was frustrated and wanted nothing more than to run forward but my mind had no choice but to walk, a little slower . . . a little slower maybe, a little different but I finally got to the place I wanted to be.

Landmark West, I thank you for giving me the courage and the willingness to try through so many years of struggle.

To my friends, I thank you. To anyone who has ever shared the joy and the pain, to anyone who has ever gotten lost within the laughter as well as the chaos, to those of you who have ever left footprints in my heart and sweet memories in my head: I will never forget you.

To my family, who let me be exactly what I am, your wisdom taught me strength and your love taught me compassion; I love you.

Through the years, I have learned more than I could have ever learned in the classroom. I've learned to never give up on the important things, the ones that, once accomplished, leave you that feeling in yourself that you can do anything.

I've gained the power of knowing that no matter what lies ahead of me . . . is an opportunity for experience.

But most of all, I've learned that overcoming adversity, like a broken bone, causes you to heal that much stronger.

I walk away from here today a high school graduate taking with me invaluable lessons; what I leave behind is an insecure 8-year-old who couldn't read or write and believed she never would. I don't really know what lies in front of me; but whatever it is and whenever it comes I welcome it and look forward to its challenge, secure in the knowledge that I can do anything!

LECTURE 1

Larry B. Silver, M.D.
Georgetown University School of Medicine
Clinical Professor of Psychiatry
Director of Training in Child and Adolescent Psychiatry

THE ROLE OF THE FAMILY IN HELPING THE CHILD OR ADOLESCENT WITH LEARNING DISABILITIES

*"With the understanding and often not-too-gentle persistence of my teachers I worked through my anger and confusion and became confident, secure, and even diligent."**

**Note: This and all subsequent chapter opening quotes are from individuals with learning disabilities.*

Introduction by Richard L. Goldman

I would like to welcome everyone to the first installment of the Landmark West Speaker Series. Landmark's mission is to educate the dyslexic child and to help the dyslexic cope with his or her learning disability. In addition, the school attempts to transcend the daily education of 170 students, through outreach and public awareness. Our Speaker Series helps meet this goal as experts discuss family, social and educational issues related to Learning Disabilities.

To inaugurate our series, we are fortunate to have Larry B. Silver, M.D., an internationally recognized authority on ADHD and children with learning and emotional problems. A noted author and lecturer, Dr, Silver is Clinical Professor of Psychiatry and Director of Training in Child and Adolescent Psychiatry at Georgetown University School of Medicine. He has served as Acting Director and Deputy Director of the National Institute of Mental Health. He is the author of *The Misunderstood Child; Attention Deficit Hyperactivity Disorder: A Clinical Guide to Diagnosis and Treatment*, and, *Dr. Larry Silver's Advice to Parents on Attention Deficit Hyperactivity Disorder.* Dr. Silver currently is a board member of the Learning Disabilities Association (LDA) and other associations for children with learning disabilities and ADHD.

Tonight, Dr. Silver will discuss learning disabilities and attention deficit disorders as life disorders affecting youngsters through adolescence and in all aspects of their lives. He will demonstrate how the child can be successful within the family and with friends. Without further ado, it is a great pleasure to introduce Dr. Larry Silver.

Dr. Larry Silver

Tonight, I would like to talk about your roles as parents of a son or daughter with a learning disability. Your role in helping them is critical if they are to have the opportunities of becoming successful adults. School is only part of their lives. If they live in a family that does not understand, or live in a community that does not understand, their problems are magnified.

Several Landmark students that I talked to today made comments like, *"My problem at home is that my mom knows I have a learning problem, but my dad doesn't believe it"*; or, *"One of my parents is still yelling at me that if I just try harder I would do better. I wish you could help them understand I'm trying as hard as I can, but I can't do any better than I'm doing."* If parents do not understand, there is one more hurdle for the youngster to overcome.

I would like to take a few minutes to review all the terms I will use. Just this evening you heard me introduced as talking about dyslexia, learning disabilities, and ADD. I will take the first few minutes to review an historical perspective of how we got where we are. From there, we will move to some of the key issues related to learning disabilities.

HISTORICAL BACKGROUND

In the U.S. prior to 1940, if children had trouble learning, they were put into one of three major categories: (1) those children who were mentally retarded; (2) those children who had emotional problems; or (3) those children who were socially and culturally disadvantaged. By the early 1940s, we began to recognize that there was a fourth group of children who were having trouble learning because of the way their nervous systems functioned. Initially, it was thought that the reason this group of children had trouble functioning was because their brains were damaged. Yet, these children looked normal, so the term applied to this group was *Minimal Brain Damage*.

Gradually, by the late 1940s and early 1950s, more and more evidence was presented that demonstrated no damage to the brains of these children. Instead, there existed *"difficult"* wiring, or faulty neural functioning. So the name of this disorder was changed to *Minimal Brain Dysfunction*.

By the early 1960s, people were so confused about what *"Minimal Brain Dysfunction"* meant that the National Institutes of Health brought together what was called a "Consensus Conference" to summarize all of the research and conclude what was meant by this term. The panel concluded that *Minimal Brain Dysfunction* referred to a group of problems often found together, where the child had trouble learning because of the way his or her nervous system operated. Secondly, many children in this group were hyperactive and/or distractible. Third, many of them had emotional, social, and family problems.

If we had stopped there, we would have been years ahead of ourselves, but like most government documents, the "Conference" report was put on the shelf, collected dust, and was ignored. We had to go through twenty years of re-inventing the wheel before we discovered the same conclusions. That is, if we look at the children and adolescents who go to a school like Landmark, we find a common theme, that they have a learning disability. We find that about twenty to twenty-five percent of them also will have *Attention Deficit Hyperactive Disorder* (a term I'll define for you later). In addition, many of these children have social, emotional, and family problems.

What causes these secondary emotional, social, and family problems? Let me give an example. A boy begins school. He gradually falls behind academically and is kept back. He is now a year older and a head taller than everyone else in his class. He is still not learning and begins to feel bad about himself. He becomes totally discouraged and starts to misbehave in school. His teacher then calls home telling his parents, *"Your child is not doing this; your child is not doing that."* His parents begin to feel badly and become frustrated, as does the child. It's 100% predictable that if there are two parents in the family, one will believe that the best way to help the child is to be firm and strict, while the other will believe that the best way to help the child is to be understanding and permissive. And so the parents begin to clash with each other. Eventually, the principal calls the parents in and says, *"Your child is not learning due to emotional problems, obviously due to marital conflict. Go see a mental health professional."* And so these parents go to see a psychiatrist, a social worker, or a psychologist because their child is misbehaving in school. Here, everyone is looking at the smoke rather than at the fire. As such, when we talk about social, emotional, and family problems, it is critical that we determine whether the social, emotional, or family problems are causing the academic difficulty or whether the social, emotional, or family problems are consequences of the academic difficulty.

Initially, the terms that were used for these students attempted to label the presenting issue. If the problem involved reading, the child had dyslexia. If the problem involved math and calculations, the child had dyscalculia. If the problem involved written language and graphics, the child had dysgraphia.

Gradually, it became clear that these terms did not have much meaning. Dyslexia does not tell you "**why**" a child has difficulty reading but tells you that he or she cannot read. The child may have just moved from El Salvador and only knows Spanish. Reading problems may occur for any number of reasons. The experts decided that one had to clarify the specific learning difficulties that explain why a child has trouble with reading or writing or with math. The specific learning difficulty term used today is *"Learning Disabilities."* As noted earlier, some children are also hyperactive, distractible, and/or impulsive. Many different terms have been used for this behavior. The first term used in this country was *"Hyperkinetic Reaction of Childhood."* Today, we call this disorder *"Attention Deficit Hyperactivity Disorder."* This disorder is not my topic for tonight. It is important at this time to realize that *"Learning Disability"* and *"Attention Deficit Hyperactivity Disorder"* are two separate problems. The treatment for a learning disability will not cure ADHD. The treatment for *"ADHD"* will not treat the learning disability. A parent may call me and say, *"My child has ADHD, he was put on Ritalin, and he still can't read very well."* The reason is that the reading problem relates to a learning disability and is not helped with medication.

PARENT HELP

Now let me become more specific with you about what you as parents can do to help. The first thing we need to understand is that learning disabilities are *"life"* disabilities. Learning disabilities are not just school problems. The same learning disabilities that interfere with reading, writing, and arithmetic interfere with baseball, basketball, four-square, jump rope, setting the dinner table, getting dressed, keeping a room neat—in short, with every aspect of life. You may have thought of a learning disability as *"my child can't read, reads backwards or reads upside-down and that's why he goes to Landmark."* But you need to realize that his or her learning disability also explains why there are so many problems at home or with peers. Parents must broaden their thinking and realize that their son's or daughter's brain somehow functions differently, whether he or she is with family, playing with friends, or at school.

The second thing to keep in mind is that learning disabilities are a *"lifetime"* disability. The child with a learning disability will become the adolescent with the learning disability, and will become the adult with the learning disability. Parents don't like to hear this. They want to think that their child will outgrow it. If they just give the child help for a couple of years, he or she will get better. I don't say this to make parents fell upset or depressed. I say it because, if parents face reality, they have a better chance of dealing with reality. Forty percent of the children with learning disabilities inherit the disorder from their parents. It runs in families. Parents may have the same problem. They, too, may need help.

The point is that if we get these students the right help, and if we give them the right skills and strategies for learning, they can do as well as anyone else. They might need special help through high school. They might also need to go to a college that can provide help. Today, there are graduate schools and professional schools all over the country that accommodate to youngsters with learning disabilities.

So, what is your job as a parent? Schools will deal with their academic needs. Parents need to deal with their psychological, social, family, and peer needs.

THE NATURE AND SYMPTOMS OF LEARNING DISABILITIES

I want to review what learning disabilities are, focusing on the reality that learning disabilities are not just a school disability but are a life disability. This major theme is expounded on in my book, *The Misunderstood Child*. This book is written for parents and contains information about what parents can do to help their child to be successful through adolescence.

Often, when I meet with parents who have children who have been in special education programs for years, they show me a very thick file. I'll ask them to summarize the information for me. Can they give me a list of their child's learning disabilities and, equally important, can they give me a list of their child's abilities and strengths? Most parents cannot. This information is critical. The job of a parent is to learn how to build on their child's strength's rather than expose or magnify their weaknesses. The role of special education programs is to build on the strengths while helping to compensate for or overcome the weaknesses. To do this, the parents must know their child's learning abilities as well as the child's disabilities.

It is convenient to break the types of possible learning disabilities down into simple steps. The model used is a computer-based model.

The first step in learning is to bring information to the brain and record it. This is called *"input."* The second step, once the information is in, is to make sense out of it. This is referred to as *"integration."* The third step, after input and integration, is to be able to store information so it can be retrieved again. This is termed *"memory."* The last step, after the information is brought in, integrated, made sense out of, and stored, is to get the information out again, called *"output."* Thus we talk about *input disabilities, integration disabilities, memory disabilities,* and *output disabilities.*

INPUT DISABILITIES

Some children have trouble bringing information in using their eyes and recording it properly. They have *"visual input problems."* Some children have trouble bringing information in using their ears and recording it properly. We call that *"auditory input problems."* Some children may have a mixture of both. The teacher is writing on the blackboard, while talking, and the student can have trouble bringing information in both through the eyes and ears at the same time and making sense while recording it. The term we use for this central brain process of seeing or hearing or perceiving the world is *"perception."* So the term we use to distinguish this central brain process of recording something from the outside onto the brain is *"perception."* Some children will have a *"visual perception problem."* Others may have an *"auditory perception problem."*

Visual Perception

Some children or adolescents have difficulty distinguishing differences in shapes. They may confuse b's and d's and p's and 9's. They may confuse a "3", "W", "M", and an "E" where the same symbol can appear in four different positions. They may confuse a "u" and an "n" or a "6" and a "9". This problem is normal until age six. Another visual perception problem is *"visual figure-ground."* This problem refers to difficulty differentiating between the figure one is to focus on versus the entire visual field, the background. Some children, when reading, have trouble deciding what words to look at. As they read, they skip lines, or they read the same line twice. They must to go back and catch themselves. When they look up from the page, they trouble deciding where they were when they look down again. If the table is too cluttered, their eyes look at everything but the important work on the page. When sitting at the dinner table, they may have difficulty spotting the salt, if asked to pass it. The same child might not hit the nail with a hammer. Some children have trouble with visual depth perception, bringing information in through both eyes, fusing it together, and coming up with three-dimensional vision. These are the children who fall off their chairs, or who reach for a drink and misjudge it's location, or who, after cracking an egg, let it hit the table rather than the pot.

How important is visual perception in life? What does it take to catch a ball, or hit a ball, or throw a ball? The first thing one needs to catch a ball is *"visual figure-ground."* One has to look out into the field and spot the ball from an often confusing background. The second thing one must do after one spots the ball is to keep one's eyes on it. The reason why coaches yell at children to keep their eyes on the ball is that if one's eyes are on the ball, the brain can use depth perception to figure out how fast the ball is moving so that one can get to the right place and catch it. A child who skips words and lines when reading may also have a problem with baseball and basketball. He or she will get their hands up to catch a ball too soon or too late and get hit in the face. After a while, the child will just throw up his or her hands to protect the face, because he or she is afraid of getting hit. These children do not play sports well that require this kind of eye-hand coordination.

Auditory Perception

A child or adolescent might have difficulty distinguishing subtle differences in sounds. It is easier to understand the concept of subtle differences in shapes. There are twenty-six shapes in our alphabet, and ten shapes in our numerical system. But there are 44 units of sound in the English language, called phonemes. Some words sound very similar, "Blue" and "blow", "ball" and "bell", "can't" and "can." I might say, "How are you?" and a child, say eight years old, may appear to be thinking I said, "How old are you?" This child may not be paying attention because he or she misunderstands the sounds that are heard.

Some may have an *"auditory figure-ground problem:"* If there is more than one sound at any one time, will the child know which one to focus on?

Let me give an example. John was a ten year-old boy with an auditory figure-ground problem. I observed him at his home and at his school. At the home, he was sitting and watching television. His brothers were playing a game on the floor, another sound input. The window was open and the traffic constituted another sound input. His mother was in the kitchen. Suddenly, his mother called out, *"John, please come in and set the table."* That brief message lasted about three seconds. John didn't focus on it. His mother called out three or four times and finally said to me, *"See what I mean. He never pays attention to me!!!"* And then John looked up, totally surprised, with no idea why his mother was angry with him.

John was surprised because his brain was the only brain he has ever had. He doesn't know that it is different. He is trying to go through life like everyone else; but somehow he is always being yelled at for something, and he does not know why. What John's mother needed to learn was that her child had an auditory figure-ground problem. If she wanted to talk to him, she had to make eye contact first. You have to use *"visual figure-ground"* to compensate for the *"auditory figure-ground disability."*

A few days later, I observed John in school. He was sitting at his desk working. He later described that he had heard noises in his head (thinking), noises in the classroom, and noises out in the corridor. The teacher suddenly said, *"Children, it is time to do your math, open your book to Page 16 and start Problem 4."* By the time John realized that the teacher was talking, he heard *"Problem 4."* He looked around the room and saw everyone taking out his or her math books. He took out his math book and then quietly leaned over the shoulder of the child in front of him to find out what page Problem 4 was on. At that point, the teacher said, *"John, quit bothering the other children and get to work!!!"* You see what it is like to be John, or Mary, or Allison, or anyone who is constantly being yelled at? They are accused of being stupid, or lazy, or bad, or dumb, because they cannot do things as well as everyone else. They look normal and are expected to be normal. But they are not. It is so important for parents to know their sons' and daughters' strengths and weaknesses so that they do not continually frustrate themselves and their children.

Some children will have what is termed an *"auditory lag."* It takes them a fraction of a second longer to understand what they have heard. They are constantly trying to think about what they just heard and miss what comes next. Sometimes they just can't keep up and miss a piece of information. They seem not to be paying attention. In the classroom, a teacher might explain something. Then, the child raises his hand to ask a question. The teacher says, *"I've just explained that. Why don't you listen?"* These are the children who are often called *"air heads"* or *"space cadets"* because they always seem to be misunderstanding what is being said.

Sensory Integration

There are three other areas of input problems that may not impact on learning but will impact on life. These three inputs are required to know where one's body is in space and how to perform motor tasks. The first input is *"tactile perception,"* or *"touch input."* There is *"light touch"* and *"deep touch,"* or *"pressure."* You know you are sitting down because you receive pressure from some parts of your body but not from other parts.

Some children have trouble with *"touch input."* They may not like to be held or cuddled. They complain about the tag on the back of their shirt or say that their belt is too tight. They may not like shoes and socks. Occupational therapists diagnose and treat this problem.

The second of these inputs is *"proprioception."* These are the nerve endings in muscles and joints that tell us which muscle groups are relaxed to tight and where each joint is. Some children are confused by these inputs. They might have difficulty learning to use their muscles in certain patterns, called *"muscle planning."* Anything they do that requires a pattern of muscle activity is difficult. They might have a difficulty buttoning, zippering, and tying.

The third input comes from the inner ear. Our vestibular system tells us where our head is in space and where we are in relation to the ground. Balance and movement in space may be difficult if there is a *"vestibular perception difficulty."*

So far we have only discussed 25% of the possible learning disabilities, yet I hope you are beginning to understand that these are *"life"* disabilities. The same learning disabilities that cause difficulty in school interfere with sports activities, with home activities like cutting up food and getting dressed. Visual perception skills are needed in jump rope, in hopscotch, in four square, as well as in coloring and staying inside the lines and cutting and staying on the line. All these skills require eye-hand coordination. If everyone at a Boy Scouts meeting or at an Indian Princesses meeting is drawing a turkey, then cutting it out, the child with the visual perception problem cannot draw very well or cut very well, and everyone knows what experiences such children can have. They get teased and come home and say they don't want to go there anymore.

Integration

Let me ask you to do an exercise to demonstrate the concept of integration. I want each of you to print on your brain three graphic symbols: a "d", "o", and a "g". In order to make sense out of this message there are at least three things a person has to do. The first is to package those symbols in the correct order, called *"sequencing."* Were they recorded as "god", "dog", or "ogd"? The next step is to figure out what each word means, now that it is packaged correctly, called *"extraction."* For example, "the dog" and "you dog" use the same word. In one case, the word refers to an animal, and in the other case it is an insult. The third task is to put the main inputs into a concept. This is called *"organization."*

Integration disabilities refer to sequencing, extraction, and organization difficulties. Some children have a visual sequencing or extraction problem. Others might have an auditory sequencing or extraction problem.

If a child with a sequencing problem is talking, he or she might start in the middle and go to the beginning of the thought, then shift to the end. Eventually, the message comes out so that it can be understood. The segments do not flow in the right order. This child might explain something well; but when asked to write it, he or she puts everything down out of sequence. This child might try to copy something off the board, such as "21 plus 6", but it comes out "12 plus 6." He or she transposed the "21" to a "12." Some have trouble using sequences.

I recently saw a very bright high school student and suspected a learning disability. I asked him to name the months of the year. He had no problem naming January through December. I then asked him to tell me what comes after August. There was a long pause before he said, "September." I asked why the pause, and he said, "I had to start at January and work my way up." He couldn't use the sequence of months he had learned. For these children, the dictionary is difficult. They can recite the alphabet; but when using the dictionary, they have to go back if the next letter is above and below the last letter. They have to start back at "A" each time. These are the same children who hit the ball and run to third base instead of first base. Parents get angry because every time the child sets the table, he or she cannot remember where the fork goes, or the knife and spoon. This is the same child who, when younger, put on his or her pants before his or her underpants. This child might put on a shirt and then wonder what to do with his or her undershirt. He or she cannot remember the sequence of dressing.

"Extraction" difficulties result in trouble picking up the subtle meanings of words. These children do not pick up jokes or understand humor. They don't laugh when others laugh. Jokes are plays on words, and these children don't get them. They hear things literally. Idioms or puns have little meaning to them. I saw a good example of this just a few weeks ago when I was visiting a school. It was a small special education class with about ten students. One child started talking, and the teacher said, *"Class, would you please be quiet."* Another child said, *"I wasn't talking."* The teacher replied, *"I know you weren't."* This child went on: *"But you said 'class' and therefore you meant me because I am in the class and I am very angry."* Some children appear to be paranoid because they take what is said literally.

To identify an organizational problem does not require elaborate testing. Just look at their notebook, or locker, or bedroom. One can see their trouble with organization. Their notebook is a mess, and things are in the wrong place. They may not bring home what they are supposed to bring home. Even if they do their homework, they lose it or forget to turn it in. They seem to always be losing things. Their whole life is disorganized. They may also have trouble organizing time. If you say that "a book report is due in two weeks," that time is twenty years away to them; and then the night before, they panic. They cannot plan time. They have trouble organizing themselves. These behaviors make their friends upset. Parents may find themselves getting angry when helping with homework because this child demands so much of the parent's time. If a parent says, *"Go do your homework,"* it will not get done. If this parents sits down next to the child and says," *What do you have to do tonight, what do you want to do first? Do you want to do your English first? Great! Get started, and I'll come back later to check."* The homework may get done. This child needed a parent to help organize and structure the material.

Memory

We think of two kinds of memory: *"short-term"* and *"long-term." "Long-term"* memory refers to material that is stored and can be retrieved when it is needed. For most children, *"long-term"* memory is quite good. They may remember something years old, that others may have forgotten about. If they go some place once, they know how to get back to it. Some children may have trouble with *"short-term"* memory. This is memory that is being stored and can be retained briefly. One can retain information while focusing on it; but unless it is reviewed more, the information will not stay. He or she can call a phone operator and get a number with an area code. He or she can then keep these ten digits in memory while dialing the number. But, if between getting the number and dialing it, someone starts talking, this person may forget the number.

Some children with *"short-term"* memory disabilities have a *"visual short-term memory"* disability, while others have an *"auditory short-term memory"* disability. Some parents have learned at home that they cannot give their child more than one instruction at a time. When they say, *"I want you to go upstairs, brush your hair and wash your face, then come down again,"* the child will not remember all of the instructions. In a classroom, the teacher says, *"For tonight, your homework is to read Chapter 6 and answer all the questions at the end of the chapter."* This child goes home and remembers only to read Chapter 6 and does not do the answers. At school the next day, the child is accused of not doing the homework. Children with a *"short-term memory"* disability need repetition of information to retain it. They sit down at night and memorize a spelling or vocabulary list, then go to school the next day and forget what they have learned.

When they sit in class and are shown a math concept, they really understand it. Yet when they go home that night, they have forgotten how to do it. But if a parent does the first problem, bringing it back to memory, the child can do the rest. *"Short-term memory"* problems interfere in other ways. One may have to read a chapter in a book. He or she will read the first paragraph and understand it, then the second, then the third, and fourth, and fifth. When they get to the end of the chapter, they have no idea what they have read, because they have not retained it.

Some children say *"Oh, forget it,"* or, *"It is not important."* The reason might be that they have a *"short-term memory"* disability. They start to speak to you. Half-way through, they forget what they are saying. It is awkward for them to admit, *"I forgot what I was saying."* It is easier to say, *"Forget it, it's not important."*

Output

There are two ways we get information out of the brain. One way is to use words or to talk. The other way is to use muscles, as when drawing, gesturing, writing, cutting up paper. We refer to two types of language use: *"spontaneous language"* and *"demand language." "Spontaneous language"* refers to self-initiated talking. The child has the luxury of a fraction of a second in which to organize his or her thoughts, words, and speech. Some children have no trouble with this. Some children just chatter and chatter and chatter. *"Demand language"* refers to situations where the child must respond without preplanning. What do you think the story is about?" *"What's the answer to Number 6?" "Where is your sister?"* The child must organize his or her thoughts, find the right words, and speak at the same time. Some children can't do this. The same child may be fluent and chatter with friends. Yet, when asked a question, he or she delays or say, *"What? Huh? I don't know."* He or she can't find the right words. I saw a young boy the other day and asked him what he liked to do after school. He said, *"I like to go up in my room and play with cards."* I replied, *"What do you do with your cards?* He said nothing. I then asked, *"What kind of cards?* He said, *"You know, pictures of sports, you know."* He couldn't find the right words, *"baseball cards"*, and couldn't get them out. But if I held up a baseball card, and asked him what it was, he would easily respond. There is nothing wrong with his knowledge. It is just that he cannot retrieve words or organize them fast enough. A teacher might say, *"This child is passive aggressive."* When asked to clarify, the teacher will respond, *"When he wants to, he will speak in class; but when I call on him, he refuses to answer."* This difference may occur because one task involves *"spontaneous language"* and the other task requires *"demand language."*

The same problem may occur at home. When a child struggles to get his or her thoughts out with the right words, parents and siblings get frustrated with the time needed. Eventually, someone answers for the child.

Motor Output

There are two types of motor output problems: *"gross motor"* and *"fine motor skills problems." "Gross motor"* difficulties refer to coordinating groups of large muscles: arms, legs, and trunk. Fine motor refers to coordinating teams of small muscles, like the forty-some muscles in your dominant hand when writing.

The child with a *"gross motor disability"* is clumsy. He or she cannot run well and might not learn to ride a tricycle or bicycle when everyone else does. Can you imagine what it feels like to be eight or nine years old and still need training wheels on your bike?

"Fine motor problems" most commonly impact on writing. This child holds his or her pen awkwardly and writes slowly. Printing may be preferred over cursive writing. It is a laborious effort to get anything down on paper. The child will tell you, *"My hand does not work as fast as my head is thinking."* Handwriting is messy. In addition to this mechanical problem, the child might have a written language disability, manifest in difficulty getting words down on the page. He or she makes spelling, grammar, and punctuation errors. The same child who gets A's on spelling tests will misspell the word when writing it. The same child who could recite every punctuation rule cannot apply them well in writing. This disability is a problem in the classroom when copying from the blackboard or taking notes. Homework, of necessity, is usually written work. This child might resist it, because he or she cannot write fast or easily. Written language is a problem in daily life as well. How do you write notes to your friends? If you make spelling errors, your friends might laugh at you.

THE FAMILY

Let's start with chores within the family. How do parents know what chores to give their son or daughter with learning disabilities? They can use trial and error. If, however, they know this child's or adolescent's strengths and weaknesses, they can select chores that build on their strengths, rather than expose their weaknesses. For a child with *"visual perception"* and *"visual-motor problems,"* that is, he or she has difficulty when his or her eyes must tell the muscles what to do. A parent might not ask him of her to load or unload the dishwasher (unless the family uses plastic dishes). This child could, however, walk the dog, bring in the newspaper, to take out the trash. If the child has a *"short-term memory problem,"* parents need to write down the chores. "You load the dishwasher on even number days and unload the dishwasher on odd number days."

If a child or adolescent has a *"sequencing problem,"* a parent might say, *"I know that you have a sequencing problem. So I will draw a typical place stetting on a piece of paper and put it in the kitchen drawer. When it is your time to set the table, feel free to take it out and use it to help you."* What is this parent saying? *"You have a disability. I would do anything in the world to help you get rid of it. I can't, so I must help you learn to cope. I cannot excuse you from life. I am going to teach you how to succeed in life."* Parents don't dress the child with a *"sequencing problem."* They say, *"I know you have problems when dressing, deciding what to put on first and second. So, I am going to lay your clothes out in the right order. You start at the end with your Teddy Bear and work your way down."* In this way, the child gains the competence to do things well and to compensate for problems.

Parents need to know if they must get eye contact before giving instructions. If the child appears confused, parents need to be able to know if he or she is just being difficult, or having trouble because of his or her learning disabilities. Parents cannot do this without knowledge of this person's learning disabilities.

Do not excuse these children from chores. The other children will get angry at the double standards. These other children might complain, *"How come I have to make my bed and he doesn't? How come when I do something, I get in trouble. When she does the same thing, she is excused?"*

What about other activities? I mentioned earlier that children with *"visual perception problems"* might have difficulty coloring inside the lines. What happens in Sunday School, or at Indian Princesses, or at Boy Scouts? The teacher or instructor may say, *"Draw a turkey"* or *"Cut out a turkey."* For this child, it is really going to be a turkey. The other children will laugh, and this child will not want to go any more. Suppose the parent worked with the leader, explaining that his or her child had good gross motor skills and speaks well, but has difficulty with fine motor tasks. The leader might assign this child the task of explaining the meaning of turkey and how it became the symbol of Thanksgiving. Or the leader might ask this child to hand out the glue and markers or tack up the pictures the other children have drawn. In this way, this child will be just as busy and as active as the others but will be doing things with success rather than failure. He or she will want to return. Remind the leader not to ask this child to demonstrate knot tying. It might be a disaster. Let him or her walk carrying the flag. Build on strengths rather than expose weaknesses. Create successes rather than failures.

What about camps, day camps, sleep-away camps? Let me use the same child with *"visual perception"* and *"visual-motor problems."* You know what the *"All-American Jock Camp"* is like? At the end of the week, one team feels great, while the other teams feel bad because they did not win. If this child is sent to this kind of camp, and he or she drops the ball or plays poorly, it is not hard to predict the outcome with peers. This child has *"good gross motor skills"* and can do other other activities well. Perhaps success will be with soccer, bowling, horseback riding, golf, certain track and field events, or swimming. Pick a camp that offers these types of sports. Maybe a waterfront camp with swimming, rowing, canoeing, sailing—all *"gross motor"* sports.

If a child has *"auditory perception problems,"* the coach needs to be told that this child may appear to not understand. Ask that the coach review the instructions again. Parents need to run interference. They need to help find activities at which the child can be successful. Today, at Landmark, one sixth grade girl said, *"I wish I could talk to my horseback-riding teacher."* I asked why. She responded, *"One of the things she is teaching me is balance, letting go of the reins, putting your hands on your hips and letting the horse walk. I am always afraid I will fall; but when I reach over to hold onto the saddle the instructor yells at me to put my hands back on my hips."* Someone need to explain to this riding instructor why this girl has problems with balance (maybe a *"sensory integration problem"*). Let her learn to ride and to have fun with her friends.

If a child has a *"demand language disability,"* he or she may wish to get into drama. Many might say, *"Drama . . . that is talking, and the child has a disability in this area."* The nice thing about drama class is that once the child memorizes the script, language is spontaneous. That is why many *"dyslexics"* are excellent actresses and actors. You see them on talk shows, and they may have difficulty putting two thoughts together. However, give them a script to memorize, and they win Oscars.

As children move toward adolescence, they need to begin to learn to be their own advocate. They cannot learn self-advocacy during high school. They will not be able to succeed during their post-high school years. We often make these children passive. They go to school, and the special education teacher says, *"Sit down and get this work done."* They come home, and the parents say, *"Sit down and do your homework."* Tutors often give work without explaining why. We need to help them understand themselves. In early adolescence, we have to teach them to be their own advocates. They have to know their nervous systems and their limitations and strengths and so do the parents and the professionals working with them. They may need help in talking to their friends or explaining themselves to their friends. For example, let's think of a high school student who is still reading at the third grade level and who is going to a place like Landmark. He or she may go out with friends and be handed a movie guide. Someone asks, *"What movie do you want to see tonight?"* Or, he or she is given a TV guide and asked, *"What do you want to watch?"* Reading is a problem every place he or she goes. At a restaurant, he or she cannot read the menu. These students need to know when and what to tell friends. Some of the students I met with today said that they were afraid to tell friends because *"they will think that I am stupid and they will not like me anymore."* One child said, *"I tell my friends that I go to a private school in Encino. I do not tell them it is a Landmark West because they will think I am going to a retard school."* That is how she feels about herself. We need to help students know that they are not stupid, that they are not dumb, that they are not bad, and that they are not lazy. We need to help them understand that some parts of their nervous systems are wired a little differently and, as a result, that they may need to learn differently.

We need to help them learn to be their own advocates. For example, we need to help them learn how to explain to a friend that they read slowly, or how to explain that sometimes they get lost in space, or don't know their left from their right. Parents or other adults have to role play to teach them how to explain to someone that they have a *"learning disability"* and how it interferes with their life. Parents need to be supportive when these children are working with their friends and teachers and must teach them how to fight their battles.

Let them give you an example: One mother approached the new teacher the day before school started and told her, *"My son has been mainstreamed, he has dyslexia. He doesn't read very fast, and I would appreciate it if you give him untimed tests."* If this happens, you can predict what is going to occur. First, the teacher is going to say, *"Look, Lady, I have twenty-five to thirty children in my class, five periods each day. I can't do things like that."* Or, *"Look, the child is growing up. You have to get off the child's back and let him do things."* A parent taking this kind of demanding attitude wouldn't get much cooperation. But let me tell you how we taught a child to work with his teacher. He met with the teacher himself and said, *"I want you to know that I personally picked you to be my English teacher during the 11th grade."* Don't ask your child to do this to impress the teacher. The child should know whether he or she can do better with a teacher who does a lot of lecturing, because auditory perception is a strength. Or will he or she do better with teachers who are lazy and just give out worksheets? If a child does better visually, he or she might do better with a teacher who is organized than with a teacher who is disorganized. However, the disorganized teacher may be best for some children because he or she does not teach well and does not lecture rapidly. This child can take reasonable notes from this kind of teacher. What parents should say to this student is, When you are finishing the 11th grade, talk to your friends about the teachers in 12th grade and select the teachers who best meet your needs.

The second thing this particular boy said to the teacher was, *"I know we have a lot of reading. I don't read fast, but I got the reading list in April, and I have started over the summer. I have finished reading several books. One of them is difficult, but I have arranged too have it on tape. So, I will be in good shape."* What he is saying to the teacher is, *"I have a disability, I know it will take a lot of effort on my part to do well, but I am willing to put in that effort. I am now asking you to excuse me."* It is true that a child with a learning disability has to learn differently. It is also true that the effort will take much more of his or her time. The child needs to know and accept this fact if he or she wants to be successful. The child will have to work twice as hard. He or she may have to spend twice as much time on homework.

The boy went on to say, *"When I read out loud I sometimes mess up words and stumble on words. I feel kind of foolish. Would you please never call on me in class to read out loud, because I will not do very well."* The last thing he said was, *"I have a lot of good thoughts in my head, and when I talk about them, I do great, but when I try to put them on paper, it somehow just does not come out too well. So, could I please take untimed tests, if I need to I will come in early before class and work after class, and I promise not to cheat."*

First, it would be an awfully hard-headed teacher who would tell this child, *"No."* Secondly, understand the level of understanding this student has about himself. This is our goal. Self-knowledge leads to self-advocacy.

Another example: A boy wanted to get his driver's license. His learning disabilities were *"auditory perception"* and *"auditory short-term memory."* He had no trouble with the written test. However, he did have trouble with the driver's test. He slipped behind the wheel. The examiner, a man who must do the same thing every day for years, said, *"Okay, pull out and go to the nearest stop sign. Make a left, and then wait for the next set of instructions."*

The boy blew it.

I met with the family and discussed what to do. Dad said, *"Why don't you send a letter to the Department of Motor Vehicles and say that he has a learning disability?"* The boy immediately rejected the idea because no teenager wants to be different. After going through many possibilities, the father and son reached a solution. They went to the test site, where the boy interviewed people after they had taken the test. The examiner used a standard driving test, and so it was not long before the boy had the procedure written down and memorized. So when he went back the second time, it made no difference what language the examiner was speaking. The boy knew what was going to happen. He passed the test using his strategies and compensations. It is important to teach adolescents to fight their own battles rather than doing it for them. Helping them learn how to learn and how to explain their abilities to friends is important.

SUMMARY

What am I trying to say? Each of you are here because your son or daughter has a learning disability. Some of you might say, *"No, my son or daughter just needs a small school or private classes and just a little extra attention."*

Please, erase the denial. If you don't believe it, they won't believe it. Your child or adolescent is here because he or she has a learning disability. What is important is that he or she is bright and can learn. It is important for you as parents to learn your child's profile of learning disabilities and learning abilities. If you don't know this, there is no way you can creatively or correctly help them. You have to teach your child his or her abilities and disabilities. Otherwise, he or she will go through life playing a trial and error game. You need to know how to apply this knowledge in order to build on strengths rather than to expose weaknesses. As they get older, you need to teach them to do their own advocacy.

QUESTIONS AND ANSWERS

Question: What about family problems and their psychological effects?

Dr. Silver: These children obviously have social and emotional problems. Professionals should move in and work with them. The key here is for the professional to understand the abilities and disabilities of the child as well as the parent does. It is frustrating to be in therapy with someone who does not understand the relationship between emotional problems and learning disabilities, who does not understand learning disabilities. Also, if the child is with a therapist who does a lot of talking, and if they have auditory problems, and the therapist does not know it, the therapist is going to get frustrated and accuse the child of not paying attention.

Question: My child has a short-term memory problem and cannot seem to memorize what is in a book but memorize baseball cards very well. Why is this?

Dr. Silver: Motivation is part of the factor. That is where the school come in and helps. Another thing is that some types of memory work is non-threatening, especially if it is learned at their pace, like baseball statistics. But when memorizing a spelling test, your child is taking a big risk. Many of these children do quite well in non-threatening or safe areas.

Question: My daughter doesn't want anyone to know she has a learning problem. She feels that she is stupid and that something is wrong with her.

Dr. Silver: Today at school, we discussed with your youngsters the difference between being dumb and being stupid. Being dumb, I told them, means that your IQ is below average. I reassured these children that they would not be here at Landmark if that were true. On the other hand, being stupid, I told them, means that you can't do what everyone else can do. Some children will tell me, "If I'm so smart, how come I'm so stupid?" If they cannot read as well as the other children, if they cannot do math as well, if they cannot write as well, they feel stupid. That is a very subjective feeling. I can't help them to get rid of that poor self-image until they start succeeding. Only then do they start to feel better about themselves. So you have to acknowledge the difference between the two words.

Many of these children have learned the hard way that people do not understand them and their disability. They learned that in a regular school, going to the resource room got them teased because they had to go to the "retard room." They learned that they have to cover up or hide their special treatment.

All children between the ages of thirteen and fifteen can't tolerate being different. This is an age period that is horrible anyway. At this age, they don't like to be different, they don't like anyone else being different. They have their own little cliques, their own little groups, and their own little in-groups at this age. This is an age period in which many have tremendous difficulties. Learning disabled children learn from experience that they get laughed at or teased if they tell people that they have a problem. This is an age period in which it is painful to be different. So what I think you can do about your daughter is to try to be supportive and understanding. Ask the staff here to help you. If it gets to be a real problem, you should seek some expert help to see what you can do about it. Many of the children today were telling me how much energy they put into covering up their problems. They don't want people to know what their problems are. It is hard to explain their difficulties, plus people begin to think that they are dumb.

If you can't listen and talk very well, people assume that you are an "air-head" or a "space cadet". When these children struggle with the real world, it is usually easier to cover up and not be honest when dealing with the problem. Some parents have had the experience with neighbors who upon hearing that the child is in a special education class tell their children to stop playing with the child because their is something wrong with him of her. Some of you may not have had that experience; but your children have certainly had that happen to them. So I think we have to be supportive. We have to be understanding. We have to help them to find ways of feeling better about themselves. Sometimes in high school, it becomes easier because it is easier to be honest with the best friends youngsters make there. So we must also buy some time.

Question: Is there a standard form of discipline that works best?

Dr. Silver: No. The key is for both parents, if there are two parents, to agree with a single approach and to be consistent when imposing it. What I tell parents is that in addition to being children, they also have a learning disability. Thus, you also have to form parental boundaries as to what is acceptable or unacceptable behavior in the family. Try to agree upon what is acceptable behavior. Difficulties arise when the child blows up during his first minutes home from school. He or she has held in the anger and frustration all day, and then let it out at home, where it is safe. He or she may get angry with a brother or sister who may be two years younger but who has no problems and may be breezing through school with straight A's with no effort. Thus, consistency helps, and structure is helpful. If you really can't figure out what to do, you might sit down with some of the people here at Landmark and figure out what to do. If the behavior is totally unacceptable, one of the chapters in my book describes a point system and time-out system that work with some of the children. After you get unwanted behaviors under control in these ways, you can move to more casual forms of discipline.

Question: What about the frustrations parents have when helping children with their homework?

Dr. Silver: The rule of thumb is, if homework becomes a battle zone, them pull out and let the school deal with it. If youngsters refuse to do their homework, then the school will easily find out the next day when the child returns without the homework done. If you've gotten beyond that, and the child is really trying hard, then you say, "The homework is up to you. If you want any help, then I'll be glad to help you. But if you get frustrated, and your help is just not working, then what you might say is, "I really apologize, I really thought I understood your learning problems, but clearly I don't understand well enough to teach you.[1] Why don't you just skip it tonight, and I'll call your teacher tomorrow and ask him how I can help you." Then, you get someone at the school to show you that your child will learn faster if you use methods A, B, or C. Since parents don't have backgrounds in special education, send a signal to the youngster that lets him or her know that you do not know how to help and that is your problem, that you will figure it out together with the help of the teacher. But again: if homework time becomes a battle zone, pull away, and let the school give advice and directions as how to best interact with your child during homework time.

Question: It is true that these children do not get rid of their problems? Isn't there some exercise or program or something that helps them get rid of their problems?

Dr. Silver: How many of you as adults have learning disabilities? (Hands raised). How many of you have gotten rid of them? (Laughter.) I don't say this to make you feel badly, I say this to be realistic. We do not know yet how to get rid of learning disabilities. An awful lot of people are making millions of dollars promising patients magic cures with brain studies and vibrating beds and such. These "cures" do not work. What does work is if we can teach your sons or daughters how to use their strengths and how to compensate for their weaknesses. If we can teach them strategies for learning, then they can be successful people.

My reading skills are still quite low. I was able to read a good book on the plane on the way out here because I had it on tape. My spelling is very poor, and my writing is still very slow and tedious. I use a word processor if I am to do any writing. But my listening and talking skills are quite good, so I have ended up in a field like psychiatry, versus surgery. Likewise, these children will learn how to build on their strengths and compensate for their weaknesses. Some learning disabilities can be compensated for, after which they no longer present major difficulties. Most of the problems, however, will remain to be dealt with in some way.

Question: What about social problems my child has with teachers and with other students?

Dr. Silver: One of the social problems children with learning disabilities have is that they do not read social cues. We don't know if that is a perceptual problem. They do not read that look on your face, or your body language correctly that says, "Hey, you are going too far, you are annoying me and other people." Most children by the age of two, when they are playing outside, know by your tone of voice that they had better come in this time when you call, or they will get into trouble. Most children by the age of three know by your body language when you come home from work that they can be a pest, or that they should leave you alone. Children with learning disabilities don't know the meaning of these cues. They just blunder into social errors. We can teach them how to read social cues. We teach it through social skills training, and we teach it much as we do reading. We break reading down into steps, and then we put it back together again. We do the same thing when teaching social skills and the interpretation of social cues.

I am teaching social skills to some nine year-olds now. Last week, we worked on how you ask someone what time it is. This sounds very obvious. However, when you are on the phone, and the child is asking you what time it is, he or she is exhibiting this problem. In the training session, we broke "asking for the time" down into steps. Step number one was, "How do you know who may be able to tell you?" (You look to see if they have a watch on their wrist.) Step number two was, "You walk up to them." Step number three was, "You say, 'Excuse me.'" Step number four was to ask, "Can you tell me what time it is?" Step number five was to say, "Thank you." As the children role-played those skills, they began to understand. You have to teach these skills. Sometimes you have to teach them through exercise intended to help them find out what different kinds of facial expressions mean, what various kinds of voices mean, and why some people get irritated when you do not read their non-verbal cues accurately.

Question: Could you talk about diagnosis. I have a child who is a borderline LD child. But it is clear that the teachers feel she has definite problems, and this is impacting upon our family.

Dr. Silver: I will stick with learning disabilities as a generic term, rather than dyslexia, which is one aspect of learning disabilities. There is a difference between having a learning disability and being eligible for services. Having a learning disability means that there is evidence that your child is learning differently in certain areas. The way you sort that out is to use one or more of three sets of tests. The first test is an intellectual assessment. This doesn't have to be an IQ test, but you need to know what your son's or daughter's intellectual potential is and whether they are overachieving or underachieving. Also, you should look at various measures of intelligence and look for any consistencies. Scores I saw yesterday included a verbal score of 148 and a performance score of 96, a tremendous difference. The second test involves some sort of achievement test. The achievement test gives a feeling for the discrepancies between performance and potentials. If there is enough of a discrepancy, a third set of test may be used that specifically recognize and diagnose learning disabilities. The most popular test is the Woodcock-Johnson, but there are many others around.

These three sets of tests, a psycho-educational battery, will give you a diagnosis. The school system will go one step further. Under Federal law, there are discrepancy formulas that tell if the child falls far enough behind to qualify for services. Depending on the budget and how many children they are trying to service, the school district may change the equation to decrease numbers coming into the system or to increase children coming out of the system. School district personnel might sit down with you and say, "Yes, I agree that your child has a learning disability, and I agree with all these scores; but he or she is not two years behind in a skill area. Therefore, he or she is not eligible for services. Or, they may agree with you that your child has an IQ of 140, so getting average grades is underachieving of failing. However, they may not pay attention to the fact that you are doing two to three hours of homework per night with your child. What you are doing is teaching your youngster and doing the work for him or her. The only way for your child to be eligible for services is for you to stop helping, to pull the rug out from under him or her, and to permit your child to fail. Then, school personnel will say, "Yes," and extend help by permitting him or her to obtain services for learning disabled children. You may decide not to do this, as it would not be fair to your child.

The definition of a learning disability is reasonably clear, the tests needed to make the diagnosis are reasonably clear, but school systems will use different scores and formulas and will debate with you when deciding whether or not your child needs and will receive services. To expand that answer and to make it specific to your child, you will need to sit down with some staff here at Landmark and go over the issues.

Question: My child startles easily, and I want to know whether this is indicative of my problem.

Dr. Silver: How old is she?

Parent: Fifteen; but she has always been that way.

Dr. Silver: In general, if something is going to impact upon the nervous system (in the form of trauma, head injuries due to auto or motorcycle accidents, trauma during delivery or brain surgery), usually problems develop that reflect the specific area of the brain that is affected. If your child's nervous system is impacted upon during pregnancy, then any are of the brain that is developing at that point is vulnerable. That is why many of your children have multiple sets of problems. Learning disabilities may thus be accompanied by hyperactivity, Tourette's Syndrome, or seizure disorders. Learning disabled children often have clusters of problems. A pronounced startle reflex usually means that a primitive reflex is lasting longer than would normally be expected. Another example that you may have seen when your children were young was that they had problems with a stretch reflex. Most children get rid of this problem early in life. The stretch reflex is elicited by pressing on a muscle so that it will retract. When you press on the hand, for example, the hand will grasp. In some of your children, this type of response may have lasted longer, and so what you may have discovered is when you put the child on your arm to feed and put pressure on the back muscles, the child backed off and extends away from you. You began to think that the child didn't like you.

There is something called the tonic-neck reflex that every child has and that goes away during the first weeks of life. That may have lasted longer in your children. The tonic-neck reflex means that when you turn your head, the arm toward which you are turning goes up over the head. Some mothers tell me that when they try to feed the child and turn the child's head toward the nipple, the child pushes the mother away. This action is only a sign that the tonic-neck reflex is lasting longer than it should. So that is one possibility.

The other possibility is that if they have auditory distractibility, reflected in difficulties blocking out sounds in their environment, then loud noises or sounds become very scary to them. They hate circuses, noisy birthday parties, fire drill alarms. Sudden noises can lead to startle reactions. There are many possible reasons for an unusual startle response. If the response continues until the age of fifteen, you should really try to help the adolescent cope with it.

Question: I have two children with learning disabilities and one teenager with none. Recently, one child without problems has become very critical of and cruel with the two other children, calling them names. What can I do?

Dr. Silver: Part of the problem with being fifteen is that one does not like people to be different. The other problem is that one does not understand the concept of "empathy"; one doesn't know how other people feel. You have to make an effort to have him or her stop the abusive behavior and to understand that other people's feelings are hurt by the abuse. However, if this advice fails, at some point you may have to say, "No more. Every time I hear you verbally abusing your brother or sister, you will spend three hours in your room. I am tired of explaining and being reasonable. I can't allow you to hinder your brother or sister's positive growth and development, so you must stop abusing and teasing them or suffer the penalties I have outlined."

Lecture 2

Franklin R. Manis, Ph.D.
University of Southern California
Professor of Psychology

CURRENT TRENDS IN DYSLEXIA RESEARCH

"Anyone can pursue their goals with hard work and dedication, no matter what hurdles stand in the way."

Introduction by Richard L. Goldman

Tonight's lecturer is oriented toward understanding dylexia. Our speaker is an Associate Professor of Psychology at the University of Southern California (USC). He specializes in research on learning disabilities in cognitive development. Dr. Manis has been conducting research studies at Landmark West for seven years, focusing on subtypes of dyslexia, reading problems as well as genetic issues related to the disability.

Dr. Manis will summarize the recent research in the field and its implications for parents and professionals. I am pleased to present Dr. Frank Manis.

Dr. Franklin Manis:

I first became interested in learning disabilities during graduate school. A visiting faculty member mentioned to me that he was starting a study with reading disabilities and I was intrigued. We began by tutoring children in a program in the Bloomington, Minnesota School District created by Mary Lee Enfield which involved a structured phonics approach to reading remediation. At that point I was struck by the paradox of the person who has excellent or average language comprehension, yet who, when coming to the printed page, has such difficulties. This was a scientific puzzle to me. I have spent my professional research career studying and investigating the cognitive bases for such learning disorders.

I have two general goals this evening. First, I want to summarize some of the recent research on the biological and cognitive bases of dyslexia. Secondly, I would like to discuss in some depth unanswered questions that I think researchers should themselves be answering. As I summarize the current research on dyslexia, keep in mind that these are only brief sketches. If you would like me to elaborate on any of the points I make, please save your questions for later.

The most basic question is: what is the origin of dyslexia? Recently it has become clear that at least some cases of dyslexia have a genetic basis. Work by Pennington, Smith and others at the University of Colorado seems to indicate that at least one form of dyslexia originates in a single gene located on chromosome 15. Their findings indicate that less than 20% of families may have the single gene form of dyslexia. The other 80% most likely are genetically heterogeneous, which means that there are many different genetic profiles, each of which can lead to the disorder we call dyslexia (see Pennington, 1989 for a good discussion of genetics and dyslexia). Behavior genetics researchers interested in dyslexia appear to agree that whatever the genetic basis for the disorder, its final common pathway, at the level of behavior, is a deficit in phonological coding of spoken and written language. I will have more to say about phonological coding shortly.

Another interesting program of research that has emerged within the last ten years has been based upon autopsy studies involving five male and two female patients with clear diagnoses of dyslexia. When stained sections of their brains were viewed microscopically at Harvard in Albert Galaburda's laboratory, it was found that all five males had common types of anatomical abnormalities that involved errors in the development and migration of neurons during the prenatal period. These neurons were found in improper numbers, and improper arrangements, particularly in language areas of the brain. Only one of the two female brains had these focal microscopic abnormalities involving neuron migration. However, both females and one male showed evidence of scarring of the cortex due to brain injury occurring some time prior to age two. In none of the seven cases was there clear documentation of trauma to the brain at birth or during early childhood. Galaburda also found that the language areas of the left and right hemisphere were of the same size. This symmetry occurs in only about twenty percent of non-dyslexic people's brains. Galaburda theorizes that the neurological basis for dyslexia consists of two anatomical traits: early cortical damage and lack of the normal asymmetry in the language regions of the brain. He speculated recently that the problem in dyslexia stems from a failure by the brain to eliminate the excessive numbers of neurons that are normally produced during the prenatal period. This results in a deviant *"neural architecture"* (patterns of connections among neurons). The deviant neural architecture underlies the unique pattern of language difficulties seen in dyslexics. The source of the anatomical abnormalities is unclear at present, but may be related to improper genetic instructions for both brain and immune system development (see Galaburda, 1989 for a good discussion of this).

As a result of these and of behavioral studies, dyslexia is no longer seen as a specific kind of learning disability involving reading and spelling. It is now coming to be viewed as a language problem with perhaps several aspects. Researchers are now concerning themselves with the nature of these language problems. Their work has implications for early diagnosis of dyslexia, remediation, and predictions about later development.

The most common difficulties involve processing the sounds of spoken language (phonological difficulties). These phonological problems may take several forms. One form involves a difficulty retrieving words from memory (so-called word finding problems). We see this occasionally in nearly everyone's speech when an individual pauses, momentarily unable to come up with the word he or she wanted to use. Sometimes the word is retrieved from memory a few seconds later, other times the speaker substitutes a definition or a similar word and other times the listener actually comes up with the word. Studies by Maryanne Wolf and others (Wolf, Bally and Morris, 1986) indicate that dyslexics have more difficulty than non-dyslexics with word finding, and this problem is very salient at the ages of five to seven.

Another form of phonological difficulty involves keeping an exact phonological representation of what someone is saying (or what the individual is currently reading). This is usually referred to as short-term memory or working memory. The idea is that proper comprehension of language involves temporarily storing words and phrases in short-term memory so that they can be analyzed for their meaning. If dyslexics make errors in storing exact versions of what they hear or read, or can only store a limited amount of information, this would interfere with comprehension. Isabelle Liberman, and her colleagues at Haskins Laboratories in Connecticut (Liberman and Shankweiler, 1985), have shown that second and third graders who are poor readers make sound-based errors in storing information in short-term memory, and this may interfere with their comprehension of sentences. Other studies have shown that problems with phonological encoding continue into adulthood among dyslexics, interfering with oral reading accuracy, reading fluency and, to some extent, reading comprehension (Aaron and Phillips, 1986; Bruck, 1987).

The most important type of phonological problem has to do with phonological analysis of speech. To understand this problem, you need to be reminded that spoken words consist of smaller parts known as phonemes. There are forty phonemes in English. Examples include the short a and long a sounds in *cat* and *cake*, the *th* sounds in *think* and *then*, and the consonants *b, d, f,* and *h*. Dyslexic children have extraordinary difficulty breaking spoken words down into their phonemes. They do not have as much difficulty separating words into syllables—the phonemes seem to be the real stumbling blocks. This problem manifests itself as early as age four, when most non-dyslexic children can tell you which word is the odd word out in the series (sun, sea, rag, sock) or the series (nod, red, fed, bed). This task involves separating words into the initial consonant (called the onset) and the remaining syllable (called the rime). Bryant and Bradley (1985) showed that children who were poor at this task at age four and five were more likely to become poor readers at age 7 or 8. Studies of adults with a history of dyslexia reveal continuing problems in dealing with phonemes. One way we have demonstrated this in our work is to ask people to remove a phoneme from a word and pronounce what is left. We use nonsense words to prevent them from solving it by spelling. For example, how would *sparf* sound without the *p* sound? *Sarf* is the correct answer. We have used this type of deletion test with individuals from grades two through ten, and the results reveal problems among dyslexics at every age (Manis, Szeszulski, Holt and Graves, 1990; Szeszulski and Manis, 1990). Bruck (1992) found that young adults with a history of dyslexia often functioned at a third or fourth grade level on this task. Bryant and Bradley have an excellent book summarizing their early work (1985). They argue that problems in analyzing phonemes makes it difficult for dyslexic children to learn to decode printed word., If you are not aware that *sun*, *sea*, and *sock* share an initial sound, and *rag* does not fit in the series, it will be difficult for you to learn a rule that says the letter *s* goes with the sound *s*.

Samuel Orton (1937) and generations of his followers in the educational realm knew that dyslexic children have difficulty learning to decode printed letters to sounds. A variety of structured programs have been devised to teach them *"phonics"* rules with some success. What Bradley and Bryant have shown is that direct instruction in phonemic analysis has an indirect effect on decoding. For example, they taught children to separate words into phonemes using pictures and found that their reading scores improved despite the fact that no direct instruction in reading was given. By far the best results were obtained when phoneme analysis instruction was combined with phonics training by using plastic letters that could be moved around to track the movements of phonemes (e.g., note how *s* moves in *sit*, *nest*, and *hits* (see Bryant and Bradley, 1985). A recent study by Hatcher, Hulme and Ellis (1994) confirms that a combination of phonemic awareness training and decoding training works better than either alone.

As we learn more about phonological problems, we are beginning to theorize about various sub-problems of dyslexia, and how they might be connected to one another. Many researchers are now claiming that the phonological difficulty is the core of dyslexia. One of the most interesting studies along this line is by Olson and his colleagues at Colorado (Olson, Wise, Conners and Rack, 1990), in which they obtained data from a large group of twins, at least one of whom was dyslexic. They tried to measure phonological skill using games, such as *Pig Latin*, which required children to move phonemes to the ends of words, and tests of the ability to decode nonsense words. Nonsense words cannot be read by sight, so the child must apply his/her knowledge of phonics (spelling-to-sound correspondence rules).

Olson and colleagues also created tasks which required the children to be sensitive to spellings (they called these orthographic tasks). They asked them, for example, to decide which of two similar spellings was the correct spelling of a word (e.g. rane vs. *rain*; dreem vs. *dream*). Olson found that the phonological tasks had strong genetic components. That is, if one member of a pair of identical twins was low in that task, then the other twin was likely to be high. The orthographic tasks were about equally related to genetics and to the individual child's exposure to reading materials.

Does this mean that dyslexics have an inherited phonological processing problem that cannot be overcome? Does it mean they must bypass phonological processing of printed words and try to memorize them as individual spelling patterns? That is not the way I read Olson's findings. Simply because phonological skill is related to genetics does not mean phonological skill cannot be taught. After all, the amount of muscle mass you have has a strong genetic base, but you can increase your muscle mass with certain types of exercises. In the same way, good phonics instruction can be expected to increase phonological skill. However, holding curriculum constant, children with high phonological skill are likely to advance faster. I am currently doing research on the orthographic component (the memorized spellings of individual words) at Landmark. Our initial findings are that over time, dyslexics' progress in reading and spelling is related to both increases in phonological decoding and memorized sight words. This suggests that you cannot progress without both components of word reading (Manis, Custodio and Szeszulski, 1993).

LONGITUDINAL STUDIES

More and more often these days we are seeing longitudinal studies of dyslexia. This is good news because longitudinal studies can shed light on the early forms that dyslexia takes as well as the long-term outcome. I will discuss studies that follow dyslexic children into adolescence and adulthood first, followed by innovative studies of young children who later became dyslexic. Studies by Margaret Bruck (1987), at McGill University, followed students who went to college. These showed that three or four factors were associated with success. I think this is important and interesting, because as parents, you have already worked on some of these factors. Dyslexics who went to college and who did well had a past history of effective academic remediation. Whether they had been exposed to a good reading program was more important than the reading level they had attained. Bruck argues that an effective reading program builds self-esteem. This means that even though the child is not reading at high levels in college, they have come out of the academic experience with high self-esteem.

Another factor which seemed to be important in Bruck's study was strong family support that continued through college. There were cases of dyslexics who had good scores and remediation, but when family support broke down, they were less successful. The third important factor was the presence of good academic support services in the college. Surprisingly, a lot of colleges do not do a good job of this, and when colleges do and the students took advantage of these services, it was found that success was more likely. The fourth factor was the intelligence test score of the students. We know IQ tests measure a large number of poorly defined skills, so it is difficult to say what was going on here. My guess is that bright students are better able to compensate for the basic language processing difficulties in dyslexia.

Turning to the other end of the age spectrum, a recent innovative study has explored what the early development of the dyslexic child may be like. Hollis Scarborough (q. v. Scarborough, 1990), a researcher in New Jersey, found forty families who had an older dyslexic child with a younger sibling who was approximately two and a half years of age. Of course the younger siblings had not been diagnosed as dyslexic yet. She also obtained approximately seventy families with an older child who was not dyslexic. Each of these families also had a two and a half year old child. The young children from both the dyslexic and non-dyslexic families were given a variety of tests at age 2.5, 3.5, 5.0 and 8 years.

Among the forty dyslexic families, twenty-two of the two and a half year old children were diagnosed as dyslexic at age eight (by means of IQ and reading scores). Eighteen of these children who were followed longitudinally were diagnosed as normal readers at age eight. Scarborough selected from a larger sample twenty control families without dyslexic children who were matched to the families with dyslexic children on a variety of criteria (sex of child, socioeconomic status, parent's education level, etc.).

What she found was that dyslexic children had problems learning the grammar of spoken language at age 2.5. These grammatical problems were not seen in the eighteen children with a dyslexic sibling who were themselves not diagnosed dyslexic or in the twenty control children. These problems were identified by analyzing the children's speech to their parents in the home. Children who later became dyslexic used shorter and less complex sentences. The dyslexics did not show unusual difficulty with articulation, comprehension and did not have smaller than average vocabularies.

Scarborough's finding is quite novel, because grammar difficulties are not widely cited as a major problem among dyslexics. They are usually cited as a minor problem, but not one that is debilitating. At the ages of three and a half and five the results were the same. At these ages, however, Scarborough also found that naming and rhyming difficulties were present in the children who were later diagnosed as dyslexic. So it wasn't until the ages of 3.5 and 5 that typical phonological problems were diagnosed. The grammatical and phonological difficulties found at these ages were correlated with reading skills at age 8. What this means is that grammatical difficulties at ages 2.5–5.0 and phonological difficulties at age 3.5–5.0 distinguish quite well between children with and without later diagnoses of dyslexia.

Scarborough's findings suggest several interesting hypotheses. First, phonological problems are not the only or most basic cause of later reading problems, but they are just an indicator that the dyslexic has a different developmental pattern. Second, Scarborough's data suggests that grammatical difficulties in the young child are also part of the dyslexic "developmental pattern." Finally, the most speculative hypothesis is that dyslexia may be a computational problem. That is, the language areas of the brain among dyslexics may not have enough capacity to compute rules in relationship to language. This deficiency manifests as a different problem at different stages of development. At age 2.5–3.5, the problem manifests as figuring out rules of grammar, because at that age this is the most difficult problem to master. At age five, the most complex problem is figuring out the language phonetically, and at age eight, the most difficult problem is mastering the rules and exceptions among printed words. Studies have shown that dyslexics have learned some of the sounds by then, but they still have problems figuring out which words fit the rules, and which words are exceptions (Rack, Snowling, & Olson, 1992).

Scarborough's view is that different manifestations of dyslexia are seen at different ages, and so it is premature to argue that phonological difficulties are the single core cause of dyslexia. I think this is a very interesting and innovating way of thinking about dyslexia. It suggests that dyslexia is a complex type of language processing problem which will manifest itself in different ways throughout the life-span. We are only beginning to understand what these manifestations are, and how they differ from one individual to another.

CURRENT RESEARCH INTERESTS

That leads me to my current research projects, which I will summarize quickly. With the help of Landmark West and the Torrance Unified School District, I have conducted a pilot longitudinal study that will help fill in the gaps in what we know about the development of dyslexics in the school years (e.g. age 10–16) (Manis et al., 1993). We followed up 21 dyslexic children undergoing intensive remediation (primarily at Landmark West). We found that they made almost two years of progress on a standardized test of isolated word reading. They also made adequate progress on a test of elementary decoding skills (e.g., pronounce the nonwords *nug* and *plunc*) and on Olson et al.'s (1995) test of orthographic skill (e.g., which is a word, *rane* or *rain*?). However, the dyslexics made very little progress on more demanding tests of phonological decoding (e.g., pronounce two-syllable nonwords), a test of phonological awareness (e.g., /p/ from /spärf/) and a test of irregular word (e.g., *beauty, colonel, silhouette*) spelling.

These findings coincide with Bruck's (1992) data on phonological awareness. Phonological skills in general are quite difficult and may never be fully mastered by many dyslexics. Dyslexics were able to compensate for the phonological problem by learning to recognize words visually. However, they could not perform the more complex task of spelling irregular words. Our data, and those of Bruck (1992) suggest either that much more attention needs to be paid to preventing or remediating phonological deficits, or that such deficits are very resistant to training and compensatory remediation is in order.

QUESTION AND ANSWER

Question: What careers and disciplines do dyslexics follow?

Dr. Manis: Bruck (1987) analyzed that question by reviewing four previous studies. She found a wide variety of occupations represented with the most common being some type of business. There were a few dyslexics in professional careers, such as medicine, teaching, etc.

Question: Are phonological problems genetic?

Dr. Manis: Olson's twin study, which I discussed earlier, indicates that phonological problems have a strong genetic basis. This does not mean that they can't be ameliorated (or on the other hand) exacerbated by the environment the child grows up in.

Question: Are these phonological problems synonymous with auditory processing?

Dr. Manis: I would use phonological problems more specifically, as involving the analysis of sounds of spoken words. Auditory processing sometimes means auditory comprehension, and, in fact, auditory comprehension seems to be one of the stronger areas in many dyslexics with average or above average intelligence test scores.

Question: Are there studies correlating conductive hearing loss with phonological problems in dyslexic children?

Dr. Manis: There have not been many studies of this relationship in dyslexics per se, but there have been studies, particularly in New Zealand and Australia (McGee, 1982) of other children who have conductive hearing losses, caused by such things as otitis media. This group of children certainly has more reading problems than average. However, a hearing problem at age three or four does not always translate into a reading problem at age eight. Some youngster can compensate, meaning that hearing problems are not synonymous with dyslexia. The majority of dyslexics do not appear to have conductive hearing losses (Vellutino, 1979).

Question: What about governing bodies who may be attempting to establish criteria for the identification of dyslexia?

Dr. Manis: We have to ask whether these definitions make use of current scientific knowledge about dyslexia. Most of them rely on a discrepancy between intelligence and reading and try to rule out other factors such as a history of emotional problems or lack of exposure to reading. However, all of the terms in the definition are difficult to operationalize and measure. Ultimately I think the individual family and school has to make a decision about treatment based on all of the information they have about the child's developmental history, test scores, behavior, attitude and responses to previous treatment, rather than rely on a set formula.

Question: What are your views on the formulation of precise definitions of dyslexia, such as those used by consensus in cases of youngsters with hearing or visual losses?

Dr. Manis: I think everything I've been saying today operates against a precise definition of dyslexia. First, there is the problem of where in the continuum of reading or spelling skills you want to make a cut-off. Second, there is the problem of dyslexic children having a variety of language difficulties beyond simply reading and spelling, such as deficits in phonological awareness and grammatical knowledge. These language difficulties have an uncertain causal relationship to reading, spelling and writing. Third, while the neurological and genetic studies help in one way by showing that some proportion of individuals with reading problems have a biological basis for the disorder, in another way they raise problems. There is a danger of over-generalizing the genetic and neuroanatomical findings. At the present time we don't have a test that will say with a certainty which children have a biological basis for their problem. One day we may have brain imaging tests that work along the same lines as PET scanners (which show levels of brain activation during a task). Research in this area has been underway for several years now. However, it is crucial to point out even if we were able to accurately diagnose deviant patterns of brain anatomy or function, this does not tell us what the type and severity of language disorder might be. We have to rely on behavioral tests for that judgment. Rather than a single definition of dyslexia, I think detailed descriptions for each child need to be utilized, based on that individual's areas of strength and weakness in language. E.g., one child might have a severe physiological problem, mild naming and grammatical difficulties, difficulties in oral reading and word attack skills and normal language comprehension. The label needs to be as descriptive as possible, and closely tied to areas that can be attacked in a remedial program.

Question: To what extent are scores on an IQ test influenced by dyslexia? Can they be falsely low, on a test like the WISC, because of dyslexia?

Dr. Manis: Keith Stanovich (1988) has written an interesting article in which he argues that IQ scores and reading are related bi-directionally. This means that the skills measured by IQ tests are somewhat predictive of success in reading, but over time, successful experiences in reading will exert affect on the IQ test score. For example, poor decoding skills lead one to read less often and lower exposure to printed texts means fewer opportunities to learn new vocabulary and new facts. Vocabulary and information tests figure prominently on tests like the Wechsler Intelligence Scale. Hence, dyslexics over time may show significant declines in IQ. This means that IQ scores are not falsely low, they are *really* low. We need to realize that IQ tests are measures of attained knowledge or skill that include both genetic and environmental influences. I would advocate giving tests that measure many aspects of intelligence, such as the Woodcock Johnson Revised test (Woodcock and Johnson, 1989). This would give a more complete picture of the dyslexic individual's strengths and weaknesses.

Question: Can dyslexia cause mathematics difficulties? How are the two related?

Dr. Manis: That is a little tricky to describe, because studies show that about half of children with severe reading problems have varying degrees of math problems. It is tempting to say that there is a common cause, but I suspect there are many other factors that contribute to math difficulties. One of the best studies which has attempted to break this apart was done by Bryant and Bradley (1985) in England. As I mentioned earlier, they identified the phonological factor by playing a little game, in which you have to pick the odd word out, such as *sun, see, sock* and *rag,* and then were asked to select out which does not fit. The right answer is *rag,* because the others start with "s." They also did that with rhymes, asking the children to select out rhyming words. What they found was that this type of task predicted later reading difficulties, but not arithmetic computation difficulties. This suggests to me that difficulties in math computation are related to somewhat different factors than difficulties in reading. However, reading skill is bound to be important when doing word problems, or generally when reading math texts. However, arithmetic skills, as well as geometric and algebraic concepts are probably largely independent of reading problems. So, if you have a child with both reading and math problems, I suspect that there is more than one underlying difficulty. The math difficulties might be related to visual perceptual difficulties and to visualization of shapes and how well you rotate them, as is seen in some studies (e.g., Strang and Rourke, 1985). It might be related to memory difficulties involving the storing of information of all types, not just phonetic information. Problems with attention or concentration can cause both math and reading problems, of course. Beyond that, causes of math problems have not been well studied. We know about one-tenth as much about mathematics difficulties as we do about reading difficulties.

Question: Can mathematics difficulties be caused by visual processing problems, as when trying to process information rapidly from right to left across the blackboard?

Dr. Manis: There are cases of children who have this difficulty. There has been more information coming out about mathematics lately, and about mathematics difficulties among learning disabled children. But a lot of psychologists think that visual processing problems are isolated cases, and that problems in mathematics are more often conceptual problems. For example, a child may not have learned basic arithmetic facts to an automatic level of retrieval, or not have mastered the place value concept and how to borrow or carry. By the time the child is exposed to long division, you have to apply rules at a quite rapid rate. Studies by Arthur Baroody (Baroody, 1987) at Columbia Teachers College, for example, reveal that children who have problems with math, have difficulties learning and applying those rules. Some attribute these problems to poor instruction in the American schools, and say that these problems don't appear in the Asian schools. There may be some truth in that. But I think that among severely math-disabled children, the problem at one level involves slow acquisition of mathematics rules. The underlying cause of the slow rate of rule learning is not clear at this time.

Question: What about motivational factors related to reading comprehension?

Dr. Manis: It is hard to find reliable measures of motivation. One of the ways psychologists attempt to measure this is by giving children a measure of print exposure, called the "Title Recognition Test." This is a quick measure of how much a child has read. We present the child a list of titles of books, including phony titles and real titles, and ask them how many books they recognize. Cunningham and Stanovich (1990) have shown that children's responses are related to orthographic skill (i.e., knowledge of word spellings) independently of phonological skill. We have replicated their study and found that the Title Recognition Test score is also highly predictive of reading comprehension, even when factors such as intelligence, memory and decoding are factored out. We think this means simply that children who are motivated to read more develop better comprehension skills. Of course the data are correlational, so it could be the other way around.

Question: Are you studying the structure of syntax and sentence variation?

Dr. Manis: No, we don't have a direct measure of that in our studies. The closest thing we have is a test of how well children can repeat sentences of increasing length and grammatical complexity. We are not measuring the child's own grammatical ability, as reflected in speech. A few studies have looked at grammar in poor readers (e.g., age 7–9) and found them to be somewhat below average (Vellutino, 1979). I'm sure it does make some contribution to reading comprehension, but I would think it would make an even larger contribution to writing skills.

Question: How do you test 2.5 year olds—on an individual basis?

Dr. Manis: Certainly we can test your children on an individual basis. Scarborough (1990) tried to get them to talk about pictures, and recorded their speech to parents. Recording spontaneous speech is the only way you can study language in young children. You can't give them formal test very easily. They analyzed the speech and recorded, for example, how many pronunciation errors were made, etc., and, as I said, it was found that dyslexics were not unusual in that respect. However, they were delayed in the development of grammar.

Question: How was it scored?

Dr. Manis: There are regular stages in the development of grammar, relating to the use of verb tenses such as the present progressive tense, the use of question forms such as *where* and *why*, the use of helping verbs such as *do* and *is*, etc. How far the child had progressed in these stages was rated by researchers.

Question: What are some examples of these grammatical errors that Scarborough studied?

Dr. Manis: It was not so much that they were errors, but delays. For example, when a child wants to say "I walked up the stairs," he might say "I walk up the stairs." If you see this in a 3½ year old, the child is using an immature form of grammar for his or her age.

Question: Later might we have the ability to analyze areas of the brain relative to different abilities?

Dr. Manis: Yes. There have been a few studies which find an association between certain abilities and areas of the brain that become active in the performance of those tasks (e.g., Duffy and McAnulty, 1985). The way this is done is by asking an individual to perform a fairly well understood task and then determining which part of the brain is active by means of PET scanning. However, PET Scanners are rare and expensive and as yet few of these studies have been done. They are an exciting possibility for the near future. Other studies have examined individuals with known damage to an area of the brain (q.v., Ellis, 1984). Those studies have given us some information about which areas are involved in different aspects of reading and spelling, as well as other intellectual tasks. This research is also still in its infancy, however.

Question: What specific kinds of problems have been identified in dyslexics' brains?

Dr. Manis: The PET scan studies show how the brain works while it is doing something, that is, they show hot and cold areas, the hot areas indicative of activity. Studies of dyslexics show some puzzling findings. For example, they show that the motor strip (in an area of the frontal lobe) is less active in dyslexics as they read (Duffy and McAnulty, 1985). However, nobody knows how that finding relates to reading problems. Duffy's studies also showed more widespread difficulties extending into the visual and language areas of the brain.

Question: Are motor difficulties a possible cause of dyslexia?

Dr. Manis: I don't know that they are a possible cause, but they certainly are associated with dyslexia. A study by Nicolson and Fawcett (1990) was based upon the premise that dyslexics have trouble automatizing any skill. What they did was to have children perform various physical activities on balance beams, etc., and at the same time count backwards from 100. What they found was that dyslexics did quite well on balance beams and the like but when they were asked to count backwards at the same time their performance deteriorated. What was striking was that one dyslexic subject was a pre-Olympic level gymnast, and even she did poorly on the balance beam when asked to count backwards at the same time. What these authors claim is that dyslexia is broader than just a language problem and involves difficulty taking any skill to a high level. Certainly, Dr. Cratty's work with motor skills here a Landmark reveals similar problems with motor coordination in a certain percentage of the dyslexic population.

Thank you and good night.

REFERENCES

Baroody, A.J. (1987). Children's Mathematical Thinking. New York: Columbia University Press

Bryant, P.E., & Bradley, L. (1985). Children's Reading Problems. New York: Basil Blackwell.

Bruck, M. (1987). The adult outcomes of children with learning disabilities. *Annals of Dyslxia: 37*, 252-263.

Bruck, M. (1992). Persistence of dyslexics' phonological awareness deficits. *Developmental Psychology, 28*, 874-886.

Bruck, M. (1990). Word recognition skills of adults with childhood daignoses of dyslexia. *Developmental Psychology, 26*, 439-454.

Cunningham, A.E., & Stanovich, K.E. (1991). Assessing print exposure and orthographic processing skill in children: A quick measure of reading experience. *Journal of Educational Psychology, 82*, 733-740.

Duffy, F.H., & McAnulty (1985). Brain electrical activity mapping (BEAM): The search for a physiological signature of dyslexia. In Duffy, F.H., & Geschwind, N. (Eds.), Dyslexia: A Neuroscientific Approach to Clinical Evaluation. Boston: Little, Brown and Company.

Ellis, A.W. (1984). Reading, Writing and Dyslexia: A Cognitive Analysis. Hillsdale, N.J.: Lawrence Erlbaum Associates.

Galaburda, A.M. (1989). Ordinary and extraordinary brain development: Anatomical variation in developmental dyslexia. *Annals of Dyslexia, 39,* 67–80.

Hatcher, P.J., Hulme, C., & Ellis, A.W. (1994). Ameliorating early reading failure by integrating the teaching of reading and phonological skills: The phonological linkage hypothesis. *Child Development, 65,* 41–57.

Liberman, I.Y., & Shankweiler, D. (1985). Phonology and the problems of learning to read. *Remedial and Special Education,* 6, 8–17.

Manis, F.R., Custodio, R., & Szeszulski, P.A. (1993). Development of Phonological and Orthographic Skill: A Two-Year Longitudinal Study of Dyslexic Children. *Journal of Experimental Child Psychology,* 56, 64–86.

Manis, F.R., Szeszulski, P.A., Holt, L.K., & Graves, K. (1990). Variation in component reading and spelling skills among dyslexic children and normal readers. In T. Carr, B.A. Levy (Eds.), Reading and Its Development: Component Skills Approaches (pp. 207–259). New York: Academic Press.

McGee, R.O. (1982). A Thousand New Zealand Children: Their Health and Development from Birth to Seven. Medical Research Council of New Zealand, Special Report Series No. 8.

Nicolson, R.I., & Fawctt, A.J. (1990). Automaticity: A new framework for dyslexia research? *Cognition, 35, 159–182.*

Olson, R.K., Wise, B., Conners, F., & Rack, J. (1990). Organization, heritability, and remediation of component word recognition and language skills in disabled readers. In T. Carr & B.A. Levy (Eds.), Reading and Its Development: Component Skills Approaches (pp. 261–322). New York: Academic Press.

Orton, S.T. (1937). Reading, Writing and Speech Problems In Children. London: Chapman and Hall.

Pennington, B.F. (1989). Using genetics to understand dyslexia. *Annals of Dyslexia, 39,* 81–93.

Rack, J.P., Snowling, M.J., & Olson, R.K. (1992). The nonword reading deficit in developmental dyslexia: A review. *Reading Research Quarterly, 27,* 29–53.

Strang, J.D., & Rourke, B.P. (1985). Arithmetic disability sub-types: The neuropsychological significance of specific arithmetical impairment in childhood. In B.F. Rourke (Ed.), Neuropsychology of Learning Disabilities. New York: Guilford.

Scarborough, H.S. (1990). Very early language deficits in dyslexic children. *Child development, 61,* 1728–1743.

Stanovich, K.E. (1988). Matthew effects in reading: Some consequences of individual differences in literacy. *Reading Research Quarterly, 21,* 360–406.

Szeszulski, P.A., & Manis, F.R. (1987). A comparison of word recognition processes in dyslexic and normal readers at two reading-age levels. *Journal of Experimental Child Psychology, 44,* 364–376.

Vellutino, F.R. (1979). Dyslexia: Theory and Research. Cambridge, MA: MIT Press.

Wolf, M., Bally, H., & Morris, R. (1986). Automaticity, retrieval processes, and reading: A longitudinal study in average and impaired readers. *Child Development, 57,* 988–1000.

Woodcock, R.W., & Johnson, M.B. (1989). Woodcock-Johnson Psychoeducational Battery–Revised. Allen, TX: DLM Teaching Resources.

LECTURE 3

James Gardner, Ph.D.
Clinical Psychologist

UNDERSTANDING AND HELPING LEARNING DISABLED STUDENTS TO SURVIVE AND THRIVE IN SOCIETY

"It is important for teachers, parents, and other students to understand the real struggles of the learning disabled."

Introduction by Richard L. Goldman

Welcome to the Speaker Series. Tonight's speaker specializes in diagnostic and psychotherapeutic work with children and adolescents, and also with the counseling of parents. Dr. Gardner currently maintains a practice in Westwood, California. His background includes direction of the Children's Center in Venice, California, Chief Psychologist of Children's Hospital in Los Angeles, and Director of the Psychological Center of Los Angeles. He has had wide speaking experience. I think some of you saw his book called the *Turbulent Teens* when you came in. This book is intended to help parents obtain information, receive encouragement, and get greater insights enabling them to better assist adolescents with learning disabilities.

Tonight he will discuss a very important topic, the ways that parents may understand and help learning disabled students, whom we call "the frightened survivors in society." It is my pleasure to introduce Dr. James Gardner.

Dr. James Gardner

Thank you all for coming out tonight. I think this turnout is a tribute to parents who have come out in the evening after a hard day's work. I am always impressed with parents who attempt to learn more about their children, and about how to interact with them more effectively.

I would like to thank you all for being here, and I will try to provide you with something of value in return. We will discuss learning disabilities, how they extend themselves into life disabilities, and how parents and others can help. To get us on the same wavelength, I will first describe how things are with a learning disabled child.

A few minutes ago, a lady asked me if any of my five children had been learning disabled youngsters. This brought back some of the scenes that happened with one of our children during the early grades. One of our youngsters, now a thriving businessman, was dysgraphic as a child. That is, he had difficulty with written expression. His dysgraphia was diagnosed through testing. However, some twenty three years ago, dysgraphia was not easily recognized, nor well known (and, except among experts, remains relatively unrecognized and undiagnosed today).

His mother and I would go to school and tell his teachers that he was not trying to be sloppy. We explained that he had coordination problems when writing.

Our son was an excellent reader, and was, and is, very bright. He just could not express himself clearly in written form. We finally enrolled him at the University of Southern California, Reading School. His reading, however, was not the issue. It was just that at this school, they were willing to reduce the quantity of written work expected of him. They tried to understand the boy and the problem.

I do, then, have a special slant into some of the problems learning disabled children face, and their families must face. A learning disabled child warps normal parenting. Often in the family there is a sense of grieving, feelings that something is wrong emerge, and questions are raised about whose fault it is. The family seems to have flawed goods and there is no Nordstroms to take the product back to.

Such a child presents innumerable problems that must be constantly worked with and assisted. Our family's efforts seemed to have paid off. As I mentioned, our son is now a successful, educated young man. All of the effort we put forth has been worthwhile.

THE NATURE OF LEARNING

We first need some understanding of how the learning process works before we can understand learning disabilities. Initially there is input. This may consist of visual, auditory, kinesthetic-tactile, or proprioceptive input.

Proprioceptive input or "muscle memory" is important in many ways. For example, when learning to spell, we know if a word is written five times (or more) it is more likely to be retained than if it is not written in this way. It is within the visual-perceptual aspects of learnings, however, where things can go very wrong for a learning disabled child. Visual symbols seem to twist. The learning disabled child may actually see words or letters backwards. Letters may be inverted.

Interpreting auditory input may also pose problems. For example, auditory figure-ground problems may arise. When you get to my age, and you are at a noisy restaurant, you often do not know what anyone is saying .... you say what? what?, get me to a booth please! It is not that your hearing is bad, it is just that you are are not discriminating sounds as well as you once did. Many learning disabled children have problems focusing upon important and relevant sounds, while eliminating extraneous sounds in their environment. Other auditory problems faced by learning disabled children include auditory lag, and poor auditory memory. In short, information which does not get into the central processing system clearly and efficiently, through both auditory and visual modalities, will not be understood. Thus, the second stage of learning may not take place in an efficient manner. This I will call "The Inner Box." Within "The Inner Box" is the processing, integration, and storing of information taking place. After the signals have come in, through one or more sense modalities, the information must be processed, integrated and stored. These processed involve the decoding of both visual and verbal information, the sequencing and synthesizing of information and its analysis.

During the third phase of learning, the person must do something with the information. Generally this involves either verbal or written output. Information comes back out in those two major expressive elements. A problem of oral expression is termed aphasia. If there is a written problem it is called dysgraphia. Various perceptual or learning problems may be related in some underlying way. A youngster may, and often does, have more than one kind of problem, with input, with integration or synthesis, or with output. Sometimes it is difficult to be sure of the problem. Thus precise diagnosis (through testing and observation) is critical.

Diagnosis of the kind of learning disability which any child can manifest is the core of successful planning for the remediation of the disability. We now have many useful diagnostic tools, such as the Wechsler Intelligence Scale for Children-Revised (WISC), and The Woodcock-Johnson Test of Cognitive Abilities, Revised. But it is not the test that we find most useful. An I.Q. test is only a rough screening device. If there is a problem with limited intelligence, then a child might not be appropriately placed in a school or class for learning disabilities. When evaluating learning disabled youngsters, it is often the sub-test analyses which are most useful. Inspecting the sub-test pattern, for example, we may find that that child has a sequencing problem, or that a child has problems with part-whole relationships. A part-whole relationship from an academic standpoint reflects problems with how well clumps of information go together, and form coherent wholes. Sorting out the events that happened at the Battle of Gettysburg, all fit together in sequence to make a coherent whole, is an example of this process. Thus a child with a whole-part relationship problem may understand different aspects of clumps of information, but still may not be able to string them together meaningfully as a 'whole' piece. Through effective diagnostic work we may, then, be able to identify such problems, and spot other problems involving input, integration processing and output. Such difficulties may cause problems not only academically, but also on the job, and even in sports.

If there are academic problems one attempts to locate a top-flight educational therapist, or in the extreme, enroll your child in a specialized school such as Landmark West. With a precise initial diagnosis one is attempting to determine just what is wrong and what should done.

PEER RELATIONSHIPS

The learning disabled child may also have problems with peer relationships. Here a score from a sub-test of the Wechsler may be helpful. If the child scores low on the sub-test called Picture Completion, he will tend to evidence difficulties when trying to pick up subtle cues, both academically and socially. Such children often miss a key part of the instructions given by the teacher, or remain oblivious to subtle social cues provided by their peers. Among peers these same children often are the ones who do not understand the subtle joke, or what just happened, or why Janie is angry at Betty. They miss the non-verbal cue of the raised eyebrow. They seem to just to plow ahead, remaining socially oblivious to what psychologists call incidental cues, and incidental learning. When these children move through social situations they may sense that others are unhappy with them. They may not know understand how the mood or direction of the group changes. They seem unaware of the fact that why they are talking about no longer fits the conversation. They are still trying to get their two-cents worth in a topic that has been long-abandoned by the others. They look and sound awkward, and they are. This problem invites derision directed toward them, and social alienation from others. In these ways the erosion of their self-concept begins.

SPORTS PARTICIPATION

In sports many learning disabled children are mismatched. Most children obviously like to play sports and do a pretty good job. The learning disabled child, however, may have both conceptual and perceptual problems. As I stand here, we all know that the parking lot is behind us, the office is up there, and Ventura Blvd is over there. Most youngsters on football fields, or on basketball courts, or soccer fields, know how players move around and have a "sense of the court," a sense of the playing field, and of the spatial relationships involved as they move around.

If a learning disabled youngster, however, does not have that spatial sense, someone has to help them get it. If they continue to have difficulties, they have to be in a sport in which they succeed. Sports such as swimming, bowling, bicycling, running and jumping come to mind. The learning disabled child with problems of organizing spatial relationships does not know what do to. Their coordination may be intact, the drive to succeed may be there. But they often have little notion of why and to where all they players are running. The overall conceptual organization is simply not there. So this kind of problem poses difficulties for kids with this type of learning disability. Incidentally, the diagnosis of these spatial relationships difficulties, the poor perception of subtle social cues, and organizational difficulties can come from parents, as well as from professionals. Informally, I see these difficulties among my young patients. When I say the elevator is to the left, as they leave my office, and every time they walk to the right, you as a parent can make such informal diagnoses yourselves. Your youngster may not seem to know where he or she is half of the time.

AT WORK

On jobs, children who are having trouble often evidence problems in sequencing, organizing and with writing and reading. Clearly, an individual who has trouble with reading and writing may well be expected to have job-related difficulties. However, problems with sequencing, organizing, and conceptualizing are of a different, though no less important nature. In a fast-food restaurant, for example, not reacting quickly and correctly may be disastrous. In such situations when A happens one is supposed to go to actions, B, C, and D quickly! The employees must do things quickly. They have to get the french fries, and the hamburger, write it all down and serve the total meal, and fast. Managers of fast-food chains will tell you that there are a lot of people who have trouble with that kind of processing. Some of their employees, they state, cannot make the necessary connections fast enough, and accurately enough.

IMPROVING ORGANIZING AND PROCESSING

One way in which parents may help their child speed up organization and processing is to engage in board games, including card games. With some children we teach card games, using slow moves. We begin slowly, but with practice we gradually speed up processing and reaction demands. We might teach the card game *Go Fish* to a 4th grader who is having processing and sequencing problems. *Later Crazy Eights* may be tried, and still later, *Spit*. Normally one must move quickly and make discriminations rapidly. In *Spit* someone puts down the 4, and then another player must play a 5 or a 3. With the learning disabled child you may start playing these games at one-half speed, and even this may be too rapid at first. You have to experiment to find the right starting speed, and then gradually become faster as the child can accommodate to processing demands. Parents (or others) may use more board games at slower speeds when dealing with younger children who need even more time for processing and integration. Then as I have indicated, you play the games with increasing speed, as rapidly as the child's developing skills will allow.

In a classroom if the teacher gives instructions too quickly this same child will also fall behind, just as he or she does in the games described. When the child falls behind in class he or she begins to manifest more avoidance behaviors, since there is inevitable failure and frustration present. These avoidance behaviors are strong predictors of failure, since then the child avoids practicing the very tasks on which he or she needs intensive work.

Thus parents need to think, not only about exotic training programs, but also what everyday things can be helpful for a child having processing, sequencing, and/or other perceptual dysfunctions. In other words we need to search for simple exercises that begin where the child is, in terms of ability, and then later progress by placing greater demands upon the child's processing speed, dexterity, and performance efficiency.

TUTORS

Tutors are still another tool to be employed against a young person's learning disability. Sometimes parents must be quite directive with the tutor. Parents must sometimes tell them how to tutor and what to tutor. Since most of us as parents cannot tutor our own children effectively, we must go outside the family when hiring a tutor. But the other side of the coin is that the tutor, no matter how skillful and experienced, cannot know your child as well as you do. The tutor, to be most effective, therefore, must be given additional information and instructions from you the parents. When we needed a tutor for our dysgraphic son, we found a person experienced with learning problems, though she had not dealt with dysgraphia before. So, his mother and I working with the tutor, devised a program which was very helpful. As our son got older, we found that the final and best help for him was a word processor. After observing the success of our son, using this strategy, I have recommended the use of the word processor to many dysgraphic students, with excellent results.

PARENTS

In addition to school and tutors, parents are a third major source of help. To be most helpful, however, parents often need to rethink their roles. Sometimes parents are locked into traditional roles. Parents are usually thought of as value setters, and as those who set the rules. Parents often pattern themselves after their own parents, or whomever they think are good parents, and try to be that kind of parent. But as I said earlier, a learning disabled child will often warp the best efforts at good parenting. What you thought was going to work will not prove successful: And who you thought you were, in your parental role does not always suffice when working with a learning disabled youngster.

One way to re-conceptualize your role as a parent, of a learning disabled child, is to consider yourself as a coach-interpreter type of person, an interpreter both of life and of life situations. You can never get out of your traditional parenting roles, those are locked in for most of us. However, you can also assume the role of coach/interpreter. This is a double job.

This additional parent-job is often useful because learning disabled children are often puzzled by the world around them. They may be confused about what is expected academically of them, or how they should act with peers, or how they might perform in sports.

Not all children are not having the same problems in all areas. However all learning disabled children will exhibit signs of confusion and/or avoidance in some area(s). A learning disabled child is thus often a puzzled child, a frustrated child and, too often, a defeated child. Self-esteem spirals downward. This in turn produces secondary emotional problems, over and above the primary problem—the learning disability itself.

Secondary emotional problems are often manifested differently in learning disabled boys and girls. Girls tend to become withdrawn, quiet, even depressed. In contrast, boys tend to act out. A learning disabled boy would rather be thought of as bad, rather than dumb. Anger is the cover for the shame of the learning disability.

One approach to understanding learning disabled children, particularly when their behavior appears in distorted forms, is to ask the simple questions as to whether they are either getting something or avoiding something by exhibiting such behavior. In this way one can "read" the behavior of children by evaluating the consequences of the behaviors manifested.

Depressed or angry acting-out behaviors may be less likely among children whose parents are willing to function as what I have termed "coach/interpreters." In this way the parent is translating puzzling aspects of the child's world into, hopefully, less puzzling aspects. The parent, as coach, then works with the child by formulating more effective solutions to the problems, whether these be social, academic, or those occurring in athletics. In other words, the parents interprets puzzles, and coaches toward solutions. Three things are needed to be a good coach: a game plan, a sense of humor and techniques, including the use of time lines, schedules, schedules, lists, check-off sheets, written notes, as well as other strategies.

OTHER TECHNIQUES

Walking Through

A useful technique, often used intuitively by many parents and teachers, may be termed "walking through." "Walking through" involves first modeling a job, and then doing it, modeling it again and then doing it again. This alternation of modeling and doing continues for as many learning trials as may be necessary.

One might, for example, help a child clean up his or her room, using the walking through technique. A personal example occurred when we moved to our new house years ago. Our child with a learning disability was supposed to join his brothers and sister in organizing each of their respective rooms. However when faced with this task, he became essentially non-functional. It was not that he rebelled and refused to do the task, it was that he seemed completely puzzled by it. As all the other children put things away, he sat in the middle of the room. He could not organize the room. He did not seem to know where to start.

Finally his mother and I recognized the problem, and asked his grandmother to help him. Granny went in and walked him through the job in the nicest way. She helped him organize his shirts, socks and pants. She moved him through the organizational maze slowly and patiently until the job was done. The room was then organized. All books and clothes had been put away neatly. Both she and my son were very pleased with themselves!

This is the walking-through technique. The parental coaching persists until some sort of criteria has been reached. Practical methods lead to success. There is no allowance for failure. In this case our son's room was organized. In other instances, organizational help may be needed for everything from time management, doing homework, keeping a note book, to straightening up a school locker.

Anxiety Reduction

While consulting to the University of Southern California's Reading School, and to other specialized schools, I began to perceive that common to all learning disordered children was much anxiety and the appearance of many avoidance behaviors. Anxiety and avoidance behaviors would be manifested depending on the task presented and upon the circumstances. The avoidance behaviors tended to be well-learned, quite automatic in their appearance, and very functional in the degree of psychic protection they offered the child.

Since anxiety usually was paired with these avoidance behaviors and seemed to serve to cue the appearance of such behaviors, we began to use techniques such as movement to reduce anxiety (large muscle activity tends to inhibit anxiety). We would ask them to read as they walked about the room. When we asked them to do this, they not only read better but anxiety also was lowered.

Some children with learning disabilities are similar to individuals who stutter. If you can get the stutterer to do something, such as standing on his or her head, talking from behind masks, or singing the words, then stuttering tends to decline. Similarly when you ask children to move away from a desk while reading, improvement in reading is often noted. By changing the stimulus cues (from a child crouching at the desk, in front of a book, to the child walking around the room with the book in hand, and teacher with arm across the child's shoulders) then anxiety is reduced and avoidance behaviors such as looking away, fooling around, sharpening pencil, tend to decline.

One–Minute Counselor

Another useful technique was formulated by my colleague Dr. Grayce Ransom, and me, at the University of Southern California's Reading School. This is essentially a one minute counseling technique for use by a classroom teacher, a tutor or a parent. Technically the technique is called "Academic Reorientation,"[1] but actually it is a one minute teaching method. Its implications go far beyond the classroom, for it can be used anytime and anywhere.

There are three essential elements to one minute counseling (it may take less than a minute, but longer than minute is too long). These steps include: (1) helping the child discriminate avoidance behavior she/he is manifesting. (2) pointing out the probable consequences of this behavior, and (3) offering a more appropriate alternative behavior.

This is how it works. Charlie, a learning disabled student, is behaving in a disruptive and inappropriate manner in class. Using the one-minute counseling technique the teacher (or aide) moves quitely over toward Charlie, bends down and points out what Charlie is doing (discriminating the behavior). "Charlie you are looking out the window and that is avoidance behavior. You are doing this because you feel you might not be successful with the work" (the explanation of the behavior, and optional aspect of the approach.

The teacher then describes the consequences. "Charlie if you keep doing this, then you will miss the lesson, you won't finish, and you'll be mad at yourself." This is followed by the offering of a more constructive alternative. "It would be best if you stopped looking out the window and keep your eyes on the book. Raise your hand if you need any help."

All this takes a few seconds. The interaction is unobtrusive and positive. There is no anger on the teacher's part, no punishment is given. The teacher has gone from just teaching academics to helping the child unlearn negative behaviors, the avoidance behaviors, and begin to learn and use more appropriate, task-oriented behaviors.

Let's move outside for another example. Charlie is with peers and his relationships have been shaky. He is becoming louder and more demanding and more unpleasant over some incident. Dad moves in and takes Charlie quitely to one side. "Charlie, you are talking too loud and becoming too bossy (stage #1). If you keep this up, your friends will not want to stay here and play with you. (Stage 2). I suggest that you lower your voice, be pleasant, and ask you friends what game they want to play, and then play it with them nicely. (Stage 3)." Of course at first all this is not going to work all the time, or even some of the time. But, throw enough mud against the bar and some sticks. Keep working with the child using the one minute counseling method and some will stick. Best of all, there are no negative side-effects. There is no down side risk to being decent with the child and trying to teach them strategies and effective behaviors that they need to know.

Circle Back

Paired with the one-minute counseling method is "circle back technique." To employ this method, the teacher, parent or coach, first uses the three stages of the counseling technique (discriminate, point out consequences, offer constructive alternatives). Then the teacher or adult moves away from the child momentarily. Then the adult circles back toward the child, and if the child is manifesting any part of the new, alternative behaviors, the adult reinforces the child with a touch, a murmured word, or with a meaningful glance.

In short the attempt is made to eliminate the old behaviors and to replace them with new, more positive ones. To do this, the child's initial attempts at the positive behaviors must be reinforced by the adult. Later the child will be reinforced by their success with academics, with social success with peers, or while participating in sports.

Psychotherapists working with children often use the one minute counseling framework. However, the professional often adds one more level to the technique, that is, a more thorough interpretation of why the child is doing the unwanted behavior in the first place.

If parents attempt to use complicated reasons for why the child is exhibiting avoidance such as "it is because of a conditioned fear of failure which interferes with effective learning!", then the parents may cause Charlie to throw up, or at least be turned off to their words. Thus we counsel parents to avoid voicing deeper level interpretations. Parent should stay with explaining what they see, what they think the consequences might be. Parents should not be too dire, but should rather be immediate. The child does not relate well to being told that he or she will not go to college if a behavior is continued.

But as Charlie's therapist, I can sit in my office with him and explore his overt behavior as both he and I see it, and as expressed in school reports. But in counseling I can also draw diagrams. We can practice new behaviors, on his part. We can practice all three parts of the brief counseling technique, and we can also delve a bit deeper into why the behaviors occur in the first place.

SUMMARY

A new role has been suggested for parents of learning disabled children. The role of parent-coach interpreter of the world. In general it is a three part role. The first job is to assess the positive or the negative aspects of your child, acting as a diagnostician. Then you should come up with a plan. What are the goals? How can you help? Who else should be brought into the plan?

Next ask yourself what techniques are available, and how might these be employed to teach the goals you have set? Most parents (and others) tend to be OK with the first two aspects, but fall somewhat short on deriving techniques. But techniques are fairly plentiful if some imagination is used. These can range from using different kinds of sports to employing board games, and can include the use of movies and music. Your child can be assisted in developing a full repertoire of social behaviors and task oriented behaviors, and will not have to fall back to using avoidance or other negative behaviors.

Parents may model many forms of appropriate behavior, using the walk through technique I have described. This can include using tasks at home, such as how to reorganize dishes when placing them in the dishwasher, as well as how to clear out and reorganize a garage, notebook or desk. Thus an emphasis should be placed upon organization, on how to get from here to there easily and successfully. There is also an emphasis upon helping your child to stop shooting themselves in the foot, with maladroit classroom and social behaviors.

The various techniques I have described are seen as ways of empowering and enabling parents.

Most of us are at our worst when we can think of nothing positive to do. As parents, we then may yell, shout and punish, but that doesn't offer our child much on the positive and constructive side of things.

I have known people who seemed to feel that if they just talk louder to a person who does not understand English, the person will finally "get it" in some manner. The same loud, useless talking may occur between a parent and child. Parents who have no useful techniques to use often tend to just become louder and/or more punitive. However, if parents reconceptualize their roles as parents, to the broader role of coach-interpreter-trainer-teacher-model, and then bring into play some of the techniques discussed (and others which you will create) your child will be the direct beneficiary.

Using these techniques should minimize your anger, and help you to think of teaching and learning trials. As you and your child go through a procedure one more time, and then again, low and behold things will begin to stick, and you will begin to see positive behavioral changes in yourself and your child.

Thank you.

QUESTIONS AND ANSWERS

Question: What kinds of techniques can you use to enhance the subtle nuances that children don't have? Those nuances with peers relations you have described, and things of that sort?

Dr. Gardner: This type of thing may sound more difficult than it actually is. What is a subtle nuance? What is a subtle nuance? Basically, it is a shading of a behavior that means something, and is usually important to pick up and not miss. Examples might range from picking up the difference in a Shakespeare class when asked a question about Richard III, when the instructor means the play, not the man, or using an arched eyebrow or a certain tone of voice in other contexts. But your question asked, how do you teach the ability to spot and to react to subtle nuances?

Movies can be used. While watching, for example, the Breakfast Club, you might ask questions such as: "What does that mean? Why did he do that? What do you think he feels about her? Does that gesture mean that he is going to beat someone up?" Comic strips can also be used. Since so many kids now watch MTV, that there is almost always a channel (no pun intended) to be explored. You might ask the child as you both view a program. "What is that image?, or perhaps, What is an image supposed to do?"

Question: How many times do you have to show a child how to organize a dishwasher, before you realize that the child is exploiting you, and you end up doing it. He really wanted you to do it. What do you do at that point?

Dr. Gardner: If that is what you suspect, exploitation, and most parents instincts are pretty good in this regard, then you might have to say "Organize any way you want to but the fact is that you can do it a lot faster and more efficiently if you do it this way." Actually I haven't found that exploitation or manipulation to be the important factors in these situations on the part of most children. Also, if the child is being manipulative then you have to ask, why is the child showing such behavior? Most children aren't born wanting to be manipulative, or mean to our parents. Parents are people who children love.

Thus, you have to ask yourself why is this happening? and What is going on here? Are you as a the parent setting up parameters that are teaching the child to be devious and/or manipulative? But if you feel that you are being manipulated on some task, such as organizing the dish washer, or giving homework help, then stop doing it.

Question: How do you deal with teenagers who react to guidance as criticism?

Answer: You have to ask yourself, as the "coach," what is the best psychology to approach this player with now? Don't forget you know his or her conditioning history better than does anyone else. Perhaps in the past you have leveled so much criticism that maybe you can no longer just switch over to the role of coach. The child may feel very bruised.

However, you might sit down together in a restaurant, which by its nature will likely promote civilized and low-key behavior, and become able to talk about the problem.

You might say "look I think I have been doing something wrong here. I believe I have been overly critical over the years. I did not mean to hurt you. But, perhaps I have made you overly sensitive to suggestions, and you read suggestions as criticism. I am just saying that I feel I have got something to offer, but I realize that my way is not the only way or even necessarily the best way. But let's try to work it out together." In this way, the teen is brought into the problem-solving and communication process.

However, nothing works all the time. You have to be prepared for that. We can formulate the best-laid plans for assisting a young person, and just have them simply not work. In such a case it's back to the drawing board and another attempt is made to figure it out all again until you have something that really works.

As parents, you keep trying. There's no alternative.

Question: As an educator what should I do about a child missing the social clues you talked about?

Answer: Speaking generally if the student seriously misreads another student's statement, or intent or something like that, I think it would be appropriate for you to say something like . . . "I think you misunderstood what Mary said" . . . and then let them work out their communication again.

Teacher response: What if I really don't know? What if it is about the war and I really don't know what is going to happen? I really believe it is expressing their own fear.

Dr. Gardner: These are really psychological issues as opposed to avoidance behaviors. When I find out they are afraid, or that it is a terrible misconception . . . like my mother says "this is the beginning of World War III!" Then it is appropriate to correct the balance. You might perhaps note that the Gulf Crisis is really not the beginning of World War III, although some people might think so. I would try not to undermine the parent, but there is a need for a balancing statement to the child. We adults are often the reality much of the time for children. We must feed them back a balanced reality. Children live in a magical world for many years, a world that is half logic and half fantasy. We know that the sense of reality and logic for children is different than that of adults. The change from children's thinking to more adult thinking begins to take place around puberty. Nevertheless, no one of us is a fully logical creature. But our children don't become logical by never learning more about reality and logic.

Question: How did you explain your son's learning disability to a complete stranger who knows nothing about learning disabilities, knowing how competitive parents can be?

Dr. Gardner: Regarding our own son, I explained it straight forwardly. I could not help if they (others) didn't understand every subtle nuance of learning disabilities. I thus explained that "Our son has a problem in expressive language, he has a problem in putting words out his fingers. Though there is no problem in language, he speaks, well, but his handwriting is poor."

If the stranger asks . . . "Is it brain damage? I would respond as follows. "I don't think of it as brain damage, although it used to be considered brain damage, I consider it more a deficit or deficiency in the hard wiring of the brain. It is like a TV set. In one little area all the wiring has not grown it, so we are just helping the wiring grow in."

If a stranger asked "What is caused your son's problem,?" I would answer "Nobody knows for sure. About one-third of children's learning disabilities are from birth defects, another third (or more) are caused by genetic factors, and the rest are caused by unknown factors. By the way, usually a stranger who asks this many questions is really seeking information about one or his or her own children.

Question: I have problems when placing my son in a new school, when I change schools he has no friends . . .

Dr. Gardner: This is not just a problem with learning disabled children. There are a number of children who have trouble making friends. Some psychologists think that if you haven't mastered "friend making" abilities by five to seven years, he or she will have trouble down the line. We see a lot of children who are social isolates, or even social misfits. Parents have worked very hard with such children. You have to choose their schools, camps, and other groups they might enter, with care. Pay special attention to the qualities and skills of leaders.

I have no glib or simple answers for this problem. Look for areas of common interest. Athletics is great, of course. But forcing a child with low skills into sports may do more harm than good. Art programs, Scouts, Computer Programs and others all can be useful. But there are not easy answers here. If possible try to place your child in situations in which there are more, rather than fewer, other children. (A big public school versus a small private school, a housing tract versus an isolated house in the hills). I realize this cannot always be an accomplished because of other factors.

Considering the hour, let me offer to stay and talk individually with any of you who may have further questions.

NOTE

1. Gardner, James, and Ransom, Grayce. Academic Reorientation: a counseling approach to remedial readers. *The Reading Teacher, 21*, 529–536, 1968.

LECTURE 4

Steven Forness, Ed.D.
University of California, Los Angeles
Professor of Psychiatry and Biobehavioral Sciences, School of Medicine
Principal of In-Patient School, Neuropsychiatric Institute

SOCIAL AND EMOTIONAL DIMENSIONS OF LEARNING DISABILITIES

"As an adult reporter, I cover news conferences, and even now someone will say a name that I can't quite process as fast as others; so I raise my hand and say, `Will you please spell that?'"

Introduction by Richard L. Goldman

Tonight we are really fortunate to have as our speaker Dr. Steven Forness. Dr. Forness is a professor and Director of the Interdisciplinary Training Program in Developmental Disabilities at the Neuropsychiatric Unit at the University of California at Los Angeles. He has done significant research in the field and has co-authored five books, including *The Science of Learning Disabilities* and *The Handbook of Learning Disabilities*. In addition, he has written over 100 articles in professional journals. The topic tonight, "Social and Emotional Dimensions of Learning Disabilities," is a crucial topic for all professionals and parents. Without further ado, it is my pleasure to introduce Dr. Steven Forness.

Dr. Steven Forness

Thank you, Rick. It is really my pleasure to be here tonight. The Landmark School has a reputation which is absolutely unparalleled in the field of learning disabilities. Just a couple of weeks ago, I recommended a young woman to the Landmark College back East. She is now going through the acceptance process. This was an interesting process to go through. The process was not what I am used to in recommending students to go on to college; to have to provide WISC profiles to a university's admissions officers.

But it is a pleasure to be here tonight, especially because Dr. Jack Cratty is here. I want to say that I live a mortal terror that some day I will get a call from Jack. He was on my dissertation committee some twenty-three years ago, and I live in high anxiety that one day Jack Cratty will call and say that he has found another flaw in my dissertation, they are rescinding my doctorate, and my whole career will go up in smoke.

Speaking of anxiety, we are going to talk tonight about some emotional-behavioral disorders and the overlap between those and learning disabilities. This is an interesting topic, particularly for people in public schools. This is because people in public schools like to place people with problems or disabilities in particular categories. There is only room for one problem in each category, as you might know, particularly if you have gone through the process of special education eligibility in public schools. If you have a learning disability, you can not be seriously emotionally disturbed, and if you are seriously emotionally disturbed, you cannot have a learning disability. If you have either one of those, you certainly can not have a speech and language problem. This has impeded our research in looking at the problem of what I call *"co-morbidity,"* which is the occurrence of two or more disorders in the same individual. I am going to speak of this tonight.

I would like to begin by taking a look at some of the figures in public schools, the way we identify children in the various categories, and inspect the trends in identification in the four largest categories in special education. These include learning disabilities, emotional, and behavioral disorders.

Then I would like to shift gears and talk about three psychiatric disorders that are the most common seen in child mental health clinics. I will also look at the co-morbidity of learning disabilities in these kids. This is based on a series of research studies that we have done in the areas of conduct disorders, attention deficit disorders and childhood depression.

Then I would like to look at a summary of research on *"sub-types of children with learning disabilities,"* the different underlying processing disabilities. There are about five major consensus sub-types of learning disabilities, and I would like to look at the particular social or emotional behavioral disorders that kids in each sub-type might have. Then I will talk about the treatment that might be appropriate for such disorders.

You will see that my own learning disability is an auditory processing problem; I need visual aids, so I am going to do everything with tables tonight.

The first table is constructed from about sixteen different appendices of the Thirteenth Annual Report from Congress on "Individuals With Disabilities Education Act." This is the new term for the original Public Law 94-142 (U.S. Department of Education, 1991). This table has the four largest categories of special education: *"Learning Disabilities"* is the largest category, next are *"Speech and Language Handicaps,"* then *"Mental Retardation,"* and then the category of *"Serious Emotional Disturbance."* (See Table 1.)

These data are interesting because you can see that the number of kids in each of those categories in the first column were being scored in 1976-77, just when the law was beginning to be enacted. This is, in effect, a baseline year, during which we really did not have mandated special education in any significant degree for kids across the entire country in any of these categories. The next column is the most recent year for which there are data. The data are delivered not only from each state's public schools, but also from the state hospitals, psychiatric hospitals, and private schools that receive special education money. Thus, it is a

Table 1. Number of Children Served in Four Handicapping Conditions in the United States and Its Territories over a Thirteen-Year Period.

Category	Years 1976–77	Years 1989–90	Difference	% Of Child Population	% Of School Enrollment	% In Main Stream
Learning Disabled	782,713	2,064,892	160% Gain	3.5%	4.8%	78%
Speech Impaired	1.171,378	976,186	18% Loss	1.7%	2.4%	95%
Mentally Retarded	820,290	566,150	38% Loss	1.0%	1.2%	28%
Emotionally Disturbed	245,481	382,159	38% Gain	0.7%	0.9%	44%

Percents rounded to the nearest Hyndreth. Total pupils served are 4.3 million, or approximately 9.8 percent.
Source: *Thirteenth Annual Report to Congress on Idea* (1991).

pretty complete count of all children in these major categories of special education.

Then, the next column shows you the gain and loss in each of these categories. The next two columns show kids in each category as a percentage of the overall population of children between the ages of zero and twenty-one. The next column is probably the most relevant one, because it shows you those kids as a percentage of all kids enrolled in public schools or in private schools, at public expense.

What is most interesting about this table is that learning disabilities in the public schools. That growth has become the point where nearly half of all our children in special education are now identified in the *"Learning Disability"* category.

There are about eight major categories in special education, and the four largest are on this table. The other categories, children with physical, visual, hearing and multiple handicaps, comprise fewer than ten percent of all kids in special education. Almost ten percent of total school enrollment are kids in special education. These four categories on the table comprise ninety percent of all kids in special education. As I noted, *"Learning Disabilities"* is almost as large as all the other categories put together and is now almost five percent of the total public school enrollment.

The next category, *"Speech and Language Handicaps,"* has suffered a slight decline in numbers over the years. People are not too concerned about that because they feel that such decline is a result of increasing recognition of the language basis of learning disabilities. Thus the children who might ordinarily be diagnosed as having a speech and language handicap, if it impacts their achievement, are now probably placed in the *"Learning Disability"* category. They do indeed have learning disabilities, but, before, some may have been placed in the *"Speech and Language Handicap"* category in most states.

The other category, *"Mental Retardation,"* as you can see, has declined considerably. What is interesting about this category is that many people do not think that those kids are being served in the *"Learning Disability"* category any more. They might have been at the outset. To diagnose a kid with learning disabilities in the public schools, we use a *"Learning Disability Discrepancy Formula,"* that takes into account how significantly behind mental age or IQ a child's academic achievement is. There has to be a significant discrepancy between IQ and academic achievement.

It is pretty clear from the table that a lot of kids now that are being identified in that category of *"Mental Retardation"* are probably kids with more severe mental retardation who do not have an IQ high enough to be diagnosed as *"Learning Disabled."* Those with an IQ too low to be diagnosed as having a learning disability, but too high to be diagnosed as having mental retardation, unfortunately, are probably not being served. I am only being partly facetious when I note that if we wait long enough, they may end up in the next category, *"Serious Emotional Disturbance* (SED)." That category is for children who have serious emotional/behavioral problems as a primary handicapping condition. As in all these categories, children with SED supposedly do not have other handicapping conditions. They have problems in school because they have anxiety disorders, schizophrenia, psychosis, attention deficit disorders, conduct disorders or a variety of other psychiatric or mental health diagnoses. You can see that this category is fairly small, composed of fewer than one percent of all school age kids.

If you think that only one percent of school age children have emotional/behavioral disorders severe enough to impact their educational performance, then *"Have we got a bridge to sell you."* Most people think that, at a minimum there should be probably two percent of kids in that

Table 2. Comparison of Changes in Children Served Between 1976–1977 and 1989–90 in California and the United States.*

Category	U.S.	California	Currently Served in California	% CA School Enroll	% In Mainstream
Learning Disabled	160% Gain	235% Gain	246,619	4.7%	71%
Speech Impaired	18% Loss	14% Loss	94,355	1.8%	94%
Mentally Retarded	38% Loss	38% Loss	24,355	0.4%	5%
Emotionally Disturbed	38% Gain	45% Loss	12,032	0.2%	10%

Percents rounded to the nearest Hyndreth.
Source: *Thirteenth Annual Report to Congress on Idea* (1991).

category. Many people also feel that in the *"Learning Disability"* category, some kids have emotional/behavioral problems that may have affected their educational performance. They are thus placed in the learning disabled category, even though they are not what we think of as having a learning disability.

The next table is even more problematic in that it gives us comparable figures for the state of California (see Table 2). California has an even greater growth in the LD category than the rest of the nation. This state is now in the bottom forty percent of all states in terms of proportion of kids in the *"Learning Disability"* category. In the *"Speech and Language"* category we have declined considerably and have substantially declined in the category of *"Mental Retardation,"* which is at four tenths of one percent of the school enrollment. In the category of *"Serious Emotional Disturbance,"* we are identifying only two tenths of one percent of school enrollment in this category of special education in California. When I say *"in special education,"* this includes mainstreamed kids who are receiving special education services in a regular class or resource room. This is not just kids in special schools, special classes and the like. Since only two tenths of one percent of kids are getting served in the SED category, we probably also have a fair number of kids with emotional/be-

Table 3. Diagnosis of Learning Disability.

Public Law 94–142 ("Individuals With Disabilities Education Act") stresses the following:

1. that a learning disability is a dysfunction in one or more of the basic *psychological processes* involved in understanding or using written or spoken language
2. that there exists a *discrepancy* between intellectual ability and attainment in reading, writing, spelling, calculating, listening or speaking
3. an *absence* of sensory or motor handicaps, mental retardation, psychosocial disadvantage, or primary emotional disturbance

havioral disorders, maybe not truly having learning disabilities, being served in the *"Learning Disability"* category in California.

The reason that I highlight this is to indicate that everything I say from here on in probably has social relevance for California.

Let us now look at the diagnostic criteria for learning disabilities in Public Law 94–142 (Table 3). You probably know this by heart if you have gone through any of the eligibility process for special education in the public schools. As you can see, this is a category that seems to stress that learning disabilities is an underlying processing dysfunction in one of the areas involved in understanding written or spoken language.

The second thing it stresses is that there is a discrepancy between IQ and academic achievement. That difference has to be in one of the major academic areas: Reading, math, and written language. As you might know, in the Federal law, if you have a disorder only in spelling, that is not a learning disability according to public school law. The reason for that is it would probably put about fifty percent of us in the *"Learning Disability"* category, me included.

The last criterion requires that we exclude from this category kids who have mental retardation, hearing handicaps, blind, visual handicaps, or physical handicaps. We also exclude kids with psychosocial disadvantage and, supposedly, kids with primary emotional disturbances. A lot of other states actually have a much higher proportion of kids in their category of *"Serious Emotional Disturbance"* because they exclude those kids from the *"Learning Disability"* category. California, however, tends not to exclude kids from the *"Learning Disability"* category if they have a primary emotional disturbance, if they meet the other criteria for learning disabilities.

Table 4. Federal Criteria for "Serious Emotional Disturbance" (SED) Diagnosis.

Classification of disorders		Limiting Conditions (all must be met)	
I,	Inability to learn that cannot be explained by intellectual, sensory, or health factors.	1.	Exists over a long period of time,
II.	Inability to build or maintain satisfactory interpersonal relationships with teacher or peers.	2.	(Exists) to a marked degree,
III.	Inappropriate types of behavior or feelings under normal circumstances		
IV.	General, pervasive mood of sadness or depression.	3.	Adversely affects educational performance,
V.	Tendency to develop physical symptoms and/or fears around personal or school problems.	4.	Must not only involve social malajustment.

The reason I mention these things to you is to give you a sense of some of the imprecision we have in defining and diagnosing learning disabilities in public schools. In effect, we do not often know, particularly in California public schools, what we have when we have a child with a diagnosis of learning disability. There is a good chance a lot of other handicapping conditions are associated with that diagnosis, some of which you would not find in the learning disability category in other states.

Now having said that, let us look at some of the criteria used to diagnose *"Serious Emotional Disturbance"* in the public schools (Table 4). These criteria are pretty much the same in California as they are in federal law. They are presented here in outline form. To qualify as having *"Serious Emotional Disturbance"* in the public schools, a child has to have a problem in one of the five areas shown on the left-hand side of the table. The full wording of the first category is: *"An inability to learn which cannot be explained by intellectual, sensory or health factors."* Doesn't that sound like a *Reader's Digest* version of a learning disability? It is. If I had to come up with a learning disability definition in a dozen words or so, that is probably the definition I would use. This is, however, the very first criterion used to qualify a child as eligible for special education in the public schools under the category of "SED."

Interestingly, the next criterion area is *"an inability to maintain satisfactory interpersonal relationships with teachers or peers."* How many of our learning disabled kids often have that problem . . . an inability to build relationships, having problems making friends with kids, having kids choose them as friends, and problems with authority figures, teachers and the like?

The next criterion is *"inappropriate types of behavior or feelings under normal circumstances."* That supposedly relates to the diagnosis of disorders such as schizophrenia in which there may be inappropriate, bizarre types of behavior. There is, however, no real definition of any of these five criteria, so we do not know if this indeed refers to the symptoms of schizophrenia.

The next one is an interesting one in that it refers to *"a general, pervasive mood of sadness or depression."* This criterion does seem to refer to the diagnosis of childhood depression as we know it in the field of psychiatry. This is, however, the only criterion of the five that seems to relate to a specific mental health diagnosis.

The last criterion seems to relate to anxiety disorders. It is *"a tendency to develop physical symptoms and/or fears around personal or school problems."* Unfortunately, we do not know for sure if the children exhibiting this have anxiety disorders. There is, as I have said, no further definition of these five criteria in federal or state laws. Therefore, we have to proceed on the basis of these symptoms however poorly defined. It is clear, however, that, if a child has a psychiatric diagnosis of depression, schizophrenia, anxiety disorders, or attention deficit disorders, such diagnoses in no way automatically qualify him/her in the SED category.

As the table indicates on the right-hand side, once a child qualifies in at least one of those five criterion areas, that problem must exist over a long period of time. It must also exist to a marked degree. In other words, it must be severe. Note that this category in the public law is called *"Serious Emotional Disturbance."* It is the only category with a word *"serious"* in front of it. We do not, for example, serve just those with severe mental retardation or just the deaf in special education, we also serve those with mild mental retardation and the hard of hearing. This is the only category with such a restriction. The third item is also a problem area. It refers to an *"adverse effect on educational performance."* This qualifier is sometimes very narrowly interpreted to mean just academic performance and not social functioning in the classroom.

And, lastly, a child can be excluded from this category if his problems can be considered only *"social maladjustment."* As with all of these other statements, *"social maladjustment"* is not further defined in the law. If you consider possible definitions or meanings for *"social maladjustment,"* I think the best definition would be almost the same as the second SED criterion, *"the inability to build and maintain satisfactory interpersonal relationships with teachers and peers."* So, on the one hand, if you have interpersonal relationship problems, you are "in"; and, on the other hand, if you have them, you are "out".

We are, unfortunately, stuck with this definition, which, as you can see in at least a couple of important respects, overlaps the *"Learning Disability"* category quite substantially, particularly with criteria one and two.

Now, having said that, let's look at three psychiatric or mental health diagnoses based on some research we have been doing at UCLA that examines this overlap with learning disabilities. The first diagnosis is that *"Attention Deficit Hyperactivity Disorder"* (ADHD). Kids with ADHD seem to have significant problems with inattention, impulsive behavior and hyperactivity. Their symptoms are that they fidget a lot, seem to run around a lot, can not seem to sit still in their seat, have problems paying attention, have a lot of problems sustaining attention and they behave in a very impulsive fashion, i.e., they seldom seem to think before they act. In the classroom, if you give them a worksheet, they will start marking the answers before they have actually read all of the questions or before they have considered all the possible choices for the correct answer. Often the kid with his hand up first in the air is also the one with the wrong answer.

These youngsters evidence a syndrome we used to know as the *"hyperkinetic child syndrome,"* or what we used to call *"hyperactivity."* It is now known as *"Attention Deficit Hyperactivity Disorder."* This is, unfortunately, a very common psychiatric diagnosis. In fact, it is probably one of the most common reasons for referrals to psychiatric clinics. It is also a diagnosis that overlaps a lot of other diagnoses, such as learning disabilities. Common estimates suggest that thirty to fifty percent of all kids with attention deficit disorders might also have a learning disability. So, youngsters with ADHD are a big share of those referred to special education classes for learning disabilities.

Further, what is interesting about ADHD is that we used to think it disappeared around the onset of adolescence, i.e., that about puberty these little hyperactive kids would lose their symptoms and no longer have ADHD. This no longer seems to be true. ADHD seems to persist well into adolescence and even adulthood. It turns out that the risk for these folks developing other psychiatric disorders or even adult criminal

Table 5. Sample of 55 Boys With Attention Deficit Hyperactivity Disorder.

Variable		ADHD		ADHD with CD/ODD	
Number of Ss		27		28	
Percentage of Minority Ss		11.1%		10.7%	
Age (years)		9.6	(1.2)	9.3	(1.1)
WISC-R:	Full Scale IQ	106.7	(13.3)	105.8	(11.1)
	Verbal IQ	106.7	(13.9)	103.7	(11.7)
	Performance IQ	105.6	(13.1)	108.7	(13.1)
	Attention Cluster	8.9	(2.6)	8.9	(2.4)
	Verbal Cluster	11.2	(2.7)	10.9	(2.2)
	Perceptual cluster	11.1	(1.4)	11.6	(2.7)
Reading Recognition Screening Test (grade level)		5.1	(2.4)	4.6	(2.1)
Reading Comprehension Screening Test (grade level)		5.2	(2.8)	4.6	(3.3)
Reading Diagnostic Test (grade level)		4.8	(2.8)	3.8	(1.7)
Math Diagnostic Test (grade level)		4.9	(2.2)	5.0	(1.7)
Number meeting modified LD discrepancy formula of one standard deviation		4		4	

behavior is really quite substantial. The follow-up studies on some of these kids into adulthood are actually quite grim. Unfortunately, it can be a tragic disorder because we do not always recognize it for what it is and get immediate help for these kids.

We studied a group of kids with ADHD, on a National Institute of Mental Health Research Grant, over the last five years (Forness, Youpa, Hanna, Cantwell & Swanson, 1982). They were carefully diagnosed as having *"Attention Deficit Hyperactivity Disorder."* We also divided them into two groups. One group had only *"Attention Deficit Hyperactivity Disorder"* and no other psychiatric disorders. Please refer to Table 5. They are in the first column on this table. Another mixed group had ADHD and also had a diagnosis of conduct disorder. Conduct disorder is a psychiatric diagnosis for kids who have very serious problems in following rules: Aggressiveness, stealing, and a variety of very serious behaviors. It is not necessarily synonymous with juvenile delinquency. It is a psychiatric diagnosis; as it turns out it also may have very poor outcomes. But both of these diagnoses, *"conduct disorder"* and *"attention deficit hyperactivity disorder,"* are diagnosed by a child psychiatrist or psychologist using DSMIII-R (American Psychiatric Association, 1987).

The diagnosis of ADHD can, however, be quite variable. To be diagnosed with *"attention deficit disorder,"* one uses a list of fourteen symptoms of which a child has to have at least eight of those symptoms. There is a lot of uncertainty in which eight need to be diagnosed. For conduct disorders, there is also a list of thirteen different symptoms, and a child has to have only three. There is variability in that diagnosis as well.

As carefully as we could, we specified five different diagnostic *"gates,"* or diagnostic reviews, that this first group in the first column had to meet in order to be diagnosed as ADHD. The group in the second column also had to meet these same criteria and, furthermore, had to meet criteria for conduct disorder. The table shows that these kids are fairly equal in age and IQ. It is interesting that their IQ profiles also matched. On their WISC profiles, both groups are actually doing pretty well on their underlying language sub-tests and in their perceptual sub-tests. Both groups, however, are doing very poorly on the sub-tests that measure attention, concentration and memory.

Now, here is what is interesting: The group that only has *"Attention Deficit Hyperactivity Disorders"* is doing pretty well academically. They are at grade level in all the academic tests we gave them, and that includes the Peabody Individual Achievement Test, the Woodcock Reading Mastery Test, and the Key Math Test. The group of kids who had attention deficit disorder and conduct disorder are academically doing very poorly, in contrast. They score a semester to a year below grade level on most academic tests. When we try to diagnose learning disabilities in both of these groups of kids by using the California formula for the learning disability discrepancy between IQ and achievement, no kids in either group qualified as having a learning disability. We then *"overrode"* the formula on the evidence that the IQ of these kids was artificially lowered because they had low subtests on the IQ having to do with their attention, concentration and memory. Our diagnosis of a learning disability could then be reasonably based on a smaller discrepancy. Even then we only got four kids in each of these groups (roughly about fifteen percent of this sample) qualifying as having a learning disability.

These results were replicated by Linda Shaywitz of Yale University

Table 6. Sample of 67 Children With Conduct Disorders.

Diagnosis	N	(%)	Age	(SD)	Male	Minority	SP .ED.
Conduct Disorders	10	(15%)	11.1	(1.6)	90%	40%	40%
Conduct disorders + Others	18	(27%)	10.0	(2.2)	94%	50%	33%
Atypical CD	21	(31%)	0.6	(2.1)	71%	40%	9%
Atypical CD + LD	18	(27%)	10.2	(2.4)	83%	33%	89%

who was one of the first, along with our group, to use a learning disability discrepancy formula to properly diagnose learning disabilities.

There is thus an overlap between learning disabilities and attention deficit hyperactivity disorder that is really much smaller than we used to think. Probably only ten to twenty percent of all kids who have attention deficit disorders have a learning disability when properly diagnosed. This is still a large number of children, because between two and five percent of the population have *"Attention Deficit Hyperactivity Disorder."* If we look at ten percent of those, we still have a fairly large percentage of kids who would be in this situation.

Let us look at another psychiatric diagnosis that is highly associated with learning disabilities, and that is the diagnosis of conduct disorders itself. We studied sixty-seven children diagnosed as having *"conduct disorder"* which, as I said before, is a cluster of severe symptoms involving aggression, rule breaking, lying, stealing and the like (Forness, Kavale, & Lopez, 1992). Their difficulties are usually diagnosed as conduct disorders only if they seem to be a relatively stable and enduring set of behaviors in a child.

In our study, we found four types of kids with conduct disorders (see Table 6). In one group, fifteen percent had only conduct disorders with no other problems. In another group, about twenty-seven percent had conduct disorders, but their primary diagnosis actually turned out to be another psychiatric disorder such as *"attention deficit disorder," "anxiety disorder"* or *"depression."* Although they had *"conduct disorder"* symptoms and could be diagnosed as having a conduct disorder, that diagnosis often masked an underlying problem such as *"depression," "anxiety disorder"* or, in some cases, even *"childhood schizophrenia."*

Almost a third (31%) had more mild forms of *"conduct disorder"* in which they were oppositional to adults, did not obey rules and gave adults all sorts of problems, but the behavior was not severe enough to qualify as having a conduct disorder in the formal sense. They were diagnosed as having either *"adjustment disorder"* or what psychiatrists call *"oppositional/defiant disorder."*

The last group (27%) had a conduct disorder but also qualified as having a learning disability. They apparently would have had a learning disability even if they did not have the conduct disorder. In essence, there were two disorders in the same child, and it was very clear after psychoeducational testing that these kids did indeed qualify as learning disabled in California, but in many other states would not qualify because they had a primary emotional disturbance. It turned out that nearly ninety percent of the kids in this last group got special education. Most of them received it through the *"learning disability"* category, not under the category of *"emotional disturbance."* Children in the other three groups had a relatively slim chance of getting into special education, particularly in California, because so few kids ever make it into that category of *"serious emotional disturbance."*

Now let me turn to what we consider a more internalizing psychiatric disorder, depression. *"Internalizing"* is a term that psychiatrists use to signify a disorder in which the symptoms are within the person, as opposed to *"externalizing,"* in which the symptoms are directed to other persons.

In *"depression"* there are four clusters of symptoms: The depressed mood, a withdrawal or inability to enjoy things, physical symptoms such as sleep disorders or loss of energy, cognitive symptoms around self-esteem or preoccupation with morbid thoughts. A child usually has to have symptoms in nearly all of these areas to qualify as having a clinical diagnosis of depression. We looked at a group of 111 kids who met all these symptoms of depression and were being diagnosed or treated in our outpatient department (Forness, 1988; Forness & Sinclair, 1990). What we found was relatively interesting (see Table 7). We found that forty two percent had only a diagnosis of *"depression"* and nothing else. We found probably another thirty four percent who actually had a diagnosis of *"depression"* but also had another diagnosis, such as conduct disorders, anxiety disorders and sometimes even attention deficit disorders.

The last group we studied had something in common with these other two groups. They were diagnosed as depressed and often diagnosed as having acting out disorders, but they also had learning disabilities. It looked as if they would indeed have had a learning disability even if they were not diagnosed as depressed. This is consistent with other articles we have reviewed that suggest that kids with depression may be at risk

Table 7. Sample of 111 Children With Depression or Dysthymia.

Diagnosis	N (%)	Age (SD)	Male	Minority	SP .ED.
Depressed or Dysthymic (only)	47 (42%)	10.2 (2.3)	67%	41%	21%
Depressed or Dysthymic (co-morbid)	38 (34%)	10.9 (2.8)	68%	24%	44%
Depressed or Dysthymic (with LD)	26 (23%)	11.4 (2.7)	62%	31%	85%

for learning disabilities (Forness, 1988; Maag & Forness, 1991). These are more than just kids who are are demoralized, sad and frustrated because they have learning disabilities. These kids have serious symptoms that are severe enough to meet criteria for a clinical diagnosis of *depression.*

You can thus see that psychiatric diagnoses may place a child at risk for having learning disabilities. Let us look now at the reverse. Do learning disabilities put one at risk for having a psychiatric diagnosis? That may actually be the case. It makes sense to a lot of teachers of kids with learning disabilities who observe emotional/behavioral problems in their kids. It also makes sense that teachers of kids with serious emotional disturbances observe that many of their kids have learning disabilities.

Table 8 summarizes literature on more than a dozen different studies on subtypes of kids with learning disabilities. In these studies, a group of kids with learning disabilities were classified by their underlying cognitive processing disorders that we referred to earlier, and these have been reviewed by Weller and Strawser (1987).

They summarized all of these studies that had been done up to that date. They also looked at the different subtypes of learning disabilities, not only in terms of the cognitive underlying processing problems that these kids have, but also at the social or emotional problems these kids have.

There are five subtypes that have been found to exist consistently across all of these studies. The first subtype is *"non-verbal organization disorders,"* what we ordinarily term *"perceptual motor problems."* These kids had learning disabilities that seemed to be the result of underlying

Table 8. Subtypes of Learning Disabilities, Classified by Their Underlying Cognitive Processing Disorders.

	Type	Problems	% of LD
1.	Nonverbal Organization Disorders	• Visual-spatial-motor deficits • Possible social misperception/ withdrawal	11–15%
2.	Verbal Organization Disorders	• Poor understanding/use of language • Possible agression/acting out	14–17%
3.	Global Disorders	• Multiple deficits in processing • Possible problems in all coping skills	8–10%
4.	Production Deficits	• Inefficient cognitive strategies • Possible inattention/ hyperavctivity	22–30%
5.	Non-LD Pattern	• Discrepancy from grade but not IQ • Possible frustration, absences	25–38%

problems in visual perception skills—visual, motor, perceptual, and spatial kinds of skills. It turns out that this group of kids within the *"learning disability"* category comprise about eleven to fifteen percent with learning disabilities, one of the smaller subgroups.

A child who has visual perceptual problems probably has problems making sense of what he sees on the page. That is why he/she may have problems in reading. They tend to reverse letters, reverse whole words, and the like. What is also interesting about these kids is that this problem may generalize to their social skills as well. When these kids have visual/perceptual problems, they also can not seem to make sense out of a social situation because of their perceptual deficits. One of the problems is *"figure/ground perception."* When I am up here talking, everybody, I am sure is not always paying attention to me. For the most part, however, you are generally able to screen out the noise of the table projector, the noise of the traffic going by, and the like. These kids can not necessarily screen out the kinds of other, distracting stimuli in the environment, particularly visual stimuli. As a consequence, when they are in social situations, they are often distracted by a lot of other things going on. They can not always tell what they are supposed to be looking at or paying attention to in social situations. They ignore some of the other important stimuli.

It is also clear that these kids tend to withdraw in social situations because they are bombarded by a lot of visual stimuli. As a consequence, they may evidence withdrawal, or even be at risk for depression. We are not certain, because this research is far too new and no one has yet examined these kids in more detail in order to arrive at a clinical diagnosis.

The next subtype is called *"verbal organization disorders."* This subtype has to do with subtle, underlying language processing problems. It is a subtype comprising about fourteen to seventeen of all kids with learning disabilities. These kids have subtle, underlying language problems that may not be severe enough to be diagnosed as a *"primary language disability,"* but that interfere with the ability to read because they do not have a substantial enough grasp of language to relate it to the visual letters on the page. Social or emotional problems can occur when these kids do not feel comfortable with the language process and get frustrated in social situations. They often act out their frustrations. These kids may thus be at risk for acting out or aggressive behaviors and be eventually diagnosed as having conduct disorders.

Not every child who has these disorders, in either language processing or visual perceptual processing, is at risk for psychiatric disorders. If they have a tendency to develop a psychiatric disorder, it might be in line with the different kinds of cognitive disabilities that occur within the *"learning disability"* category.

The next category is interesting because it is composed of children who have disorders in both of those areas. And that, thankfully, is a small subtype. Sometimes those kids are so impaired in social skills that they have pervasive problems in areas of making friends, keeping friends, knowing what to do in a social situation, and the like.

The fourth category is the largest identifiable subtype. It comprises probably twenty to thirty percent of all kids with learning disabilities and these kids have inefficient cognitive strategies. They can not seem to use their language or perception to talk or guide themselves through tasks, or through difficult academic subjects or through social situations. These are the kids who are at risk for having attention deficit disorders. They often do not have good impulse control nor good attentional skills. As a consequence, if kids have this problem, they might be at risk for a psychiatric diagnosis such as attention deficit disorders.

Lastly, the largest group are kids that, on good psychoeducational test batteries, show up as having no cognitive processing problems whatsoever. They still, however, have a discrepancy between their IQ and their academic achievement. Sometimes this is seen in kids who have high IQs and are functioning close to their respective grade levels. They are almost in the gifted range so they never get identified as learning disabled in school, or they may have low IQs and be functioning below their IQ, but not far enough below to qualify them as having learning disabilities.

Some of these will indeed find their way into the *"learning disability"* category. What is interesting about these kids, is that they comprise the largest subtype of learning disabilities in research studies to date. However, these kids do not have a demonstrateable underlying disorder in one of the processing or cognitive functioning areas, disorders they are required to have in order to qualify in the public schools as having a learning disability. Thus, many will be at risk for dropping out of school or developing a variety of social or emotional disorders.

In summary, if a child has a psychiatric diagnosis, he or she seems to be at risk for having a learning disability. If he/she has a learning disability, it does not necessarily mean he/she is going to have a psychiatric diagnosis. There may, however, be somewhere in the neighborhood of twenty-five percent of all kids in the *"learning disability"* category who might indeed turn out to have an identifiable psychiatric disorder. The vast majority may not. All psychiatric diagnoses are on a continuum. You and I know that all of our kids with learning disabilities have a variety of emotional or behavioral problems just as a result of having this learning disability. They are usually not serious enough to be diagnosed as having a clinical diagnosis and meeting criteria for mental health services. I just want to alert you that these two diagnostic categories in special education, the two systems of special education for kids with learning and with mental health problems, have a great deal in common. You need to be aware of this relationship and be on the alert. Sometimes the normal frustration and demoralization that occur with learning disabilities may spill over into having an identifiable psychiatrist diagnosis for which this child may need help from a psychiatrist, psychologist, or a mental health clinic. There are clinics, such as the NPI at UCLA, which do specialize in the emotional or behavioral disorders of children with learning disabilities and related disorders. Not all mental health clinics or practitioners may be experienced in this area and may thus fail to provide truly interdisciplinary treatment that these children require.

QUESTIONS AND ANSWERS

Question: Category 5 looks like it would be the most difficult when it comes to getting that child special education services. Could you comment on how to do that?

Dr. Forness: I assume you are talking about the fifth LD subtype rather than the fifth criterion in the SED category. It is interesting, in a mental health clinic we can give IQ tests, which often you are not allowed to give in public school. For example, L.A. city schools will no longer give IQ tests for their "learning disability" category because of a recent interpretation of an earlier class action lawsuit on minorities not being given IQ tests because such tests do not reflect their cultural backgrounds. There is a possibility that public school psychologists could use the IQ tests only to look at the underlying clusters of the IQ to determine whether or not these kids actually have an underlying processing disorder and not use the IQ itself. I do not know if it is ever even approved that only isolated subtests of the IQ could be used to show that a kid does have processing disorder and, therefore, has a learning disability, even though the use of the IQ to establish a discrepancy is proscribed. Many people are actually using processing tests that are not part of the IQ test, like the Detroit Test of Learning Aptitude, for example, the Illinois Test of Psycho-Linguistic Abilities, the Bender Visual Motor Gestalt Test, and the like, in trying to establish that maybe these kids have problems in underlying processing disorder who would thus qualify as LD. When you look at kids who do poorly on some of these tests, their performance on a full battery of tests will not really qualify them as LD. However, on an isolated test they will qualify.

In a clinic I can give a diagnosis of "learning disability" to a child who has a discrepancy between IQ and achievement and with whom I have ruled out other causes of the learning disability such as poor instruction, physical problems, and the like. Then I can diagnose a learning disability. I have to do that extra step of identifying a processing problem, however, when I go to the public schools to get that child services. Maybe about twenty or thirty percent of our kids in the L.A. public schools may not qualify as having learning disabilities because they have a discrepancy as shown on testing in a clinic but do not have both a discrepancy and a processing problem as needed in the schools. The sad news is that we may have kids with learning disabilities who qualify in our clinic, who qualify here at Landmark, but who do not qualify in the public schools of California because of the restrictive public school criteria.

Question: When you said that kids with emotional disabilities have learning problems, do you mean "cause and effect?"

Dr. Forness: Yes, the speculation is, in depression, for example, that both depression and the learning disability might be caused by a common neurologic problem (Forness & Kavale, 1985, 1991). There are some studies using sophisticated brain scans of patients with depression that suggest that they might have neurologic abnormalities. There are also studies that suggest that kids who have learning disabilities may have some neurologic abnormalities. There have been some suggestions in some of those studies, which are very preliminary, that some areas of the brain that seem to be effected in children or adults with depression may be some of the same areas of the brain that are effected in children with learning disabilities. Again, remember that depression, conduct disorders, and attention deficit disorders are all found to be associated in some children. Thus, "discipline" problems and learning disabilities could result from a common cause.

There are two other theories, in answer to your question. One is that having a learning disability puts you at risk for frustration and withdrawal. That might in turn place you at risk for depression and related conduct problems. The other theory is that if you have depression, that may put you at risk for learning disabilities because emotional/cognitive problems may make it more difficult to concentrate on academic tasks. All three of those hypotheses are still open.

Question: If you have a child diagnosed as having a learning disability and depression at the same time, and if the learning disability is treated, would that affect the depression as well?

Dr. Forness: There are a couple of answers to your question, one of which is that it is very hard, even for clinicians, to sort out the demoralization stemming from having a learning disability, versus depression from other sources. We have seen a lot of kids who come into our hospital unit that have a learning disability and depression. We have a good inpatient school for kids admitted to the psychiatric hospital. Sometimes, however, the progress they make in the classroom on their learning disability seems to result from the treatment they are receiving for depression rather than anything special that is happening in classroom remediation.

It may be that what we are doing with such a child, in relation to cognitive or social therapies or even the psychopharmacology used for depression, may also improve their general attitude, their concentration or even their neurologic processing. This, in turn, may not completely "cure" learning disabilities, but at least may help improve overall academic improvement.

This is all speculative, and there are no studies to suggest that is the case, but it is pretty clear that there are also kids in our clinic or hospital who have been treated for depression, and their depression improved; yet they still have learning disabilities.

We also have a number of studies on kids with learning disabilities who also have social skills deficits. It may be that the only reason we see a high level of social skills deficits in kids with learning disabilities is that at least some of these kids have social skills deficits, in whom the social skills problems are not due to learning disabilities per se, but are due to other types of psychiatric and behavioral disorders. We need to rethink a lot of our assumptions about kids with learning disabilities, and whether social skills are inherent in LD or are more a result of the co-occurrence of LD and psychiatric disorders.

Question: Why is attention deficit disorder part of the category of psychiatric disorders? I do not see the connection.

Dr. Forness: It is interesting that there is no diagnostic criteria for attention deficit disorders anywhere that is formally accepted, except for a few sources. The first is ICD-9, which is the International Classification of Diseases in Children. It is under the psychiatric heading in that classification. The second is DSMIII-R, which is a psychiatric diagnostic manual.

The major recognition of attention deficit disorders thus comes through the mental health system. Pediatricians also diagnose attention deficit disorders, but generally use the ICD-9 criteria. It is possible that ADD is considered a major mental health disorder not only because it is so often diagnosed by psychiatrists, but because it puts one at risk for other psychiatric disorders. There is also a shift in our thinking about psychiatric disorders recently because of recent findings in genetics, biochemistry, and neurophysiology related to psychiatric disorders; and attention deficit disorders has genetic, biochemical and neurologic features that link it with other psychiatric disorders.

Question: What steps are being taken to change the criteria in California as a result of this research that has been done to categorize these kids as "LD?" What steps are being taken in California to change what California perceives as learning disabled?

Dr. Forness: In California not much is being done at present. Nationally there are some steps being taken that I hope will impact California's recognition of these problems.

Question: Would that interest and work be under the Department of Education then? Because you said nationally it is . . .

Dr. Forness: Yes it is under the U.S. Department of Education, not under the California Department of Education. Perhaps there may be some changes made in the Federal Law definition for "serious emotional disturbance" impacting both the mental health system and the SED criteria in California. These changes involve changing SED to "emotional or behavioral disorder" (EBD), so kids who have a primary attention deficit disorder or a conduct disorder can be served.

If this happens California will feel a lot more confident in changing its learning disability definition to exclude kids who have a learning disability based on primary emotional problems, because then they would get services in the new EBD category. I am a member of the National Mental Health and Special Education Coalition, and my colleagues and I have written a new definition of emotional/behavior disorders that coalition has adopted at the federal level. It has now gone to Congress, and we are about a year away from getting that legislation passed in Congress. If we do get it passed, probably about a year or so later, California will have to change its SED criteria and broaden the scope of the kids that are served in the SED category, Hopefully, then, that will lead to a more clear cut learning disability diagnosis and will exclude kids from the learning disability category if they have primary emotional problems, because they can be served more appropriately in that second category. Also in the new SED category, we are making a provision which hopefully will change the Federal Law to make it possible for kids in any category to get mental health care related services. Right now, a child usually gets related services in mental health under Public Law 94-142 if he or she is only in the SED category. If you are in the LD category, you ordinarily do not get related mental health services funded for or provided by the school. So those would be a number of important developments that I hope will happen and would eventually affect the LD category in California.

Question: There are a lot of babies addicted to drugs in Los Angeles. They are now starting to hit the schools and there is a big concern with public schools about not being able to handle them. Probably part of the reason that they do not categorize a lot of these kids is because they can not handle them anyway. If they were identified, they do not have the teachers or the wherewithal to take care of them. But all of these kids who are coming through now who are born of mothers who were addicted to drugs, will get funding through mental health. Would they be under that already, and what is going on regarding learning disabilities with those kids?

Answer: One thing a number of folks are worried about with the drug babies is that they may end up in the learning disability category and may take slots that are available for kids who are learning disabled. What we are facing right now in California and other states is a limit on special education funds. As you know, there is a special education backlash in terms of a feeling that it is not good to be in special education, that we can serve children better in the regular classroom.

This movement is called the "Regular Education Initiative" and it seems to imply that kids with learning disabilities, emotional problems and so forth would be taken care of primarily in the regular classrooms. One of the problems is that that backlash is encouraging Congress and a lot of legislators in the various states not to provide as much money for special education as they might ordinarily give.

I see a real battle developing in regular and special education between these new populations coming into special education between those due to attention deficit disorders, prenatal substance abuse, child abuse or neglect who are needing to be served being placed in SED or learning disability categories and then there not being enough room for kids who normally would diagnosed as learning disabled.

Many people have commented on cocaine babies as a potential problem in special education. Regarding crack cocaine babies, in our program at UCLA Judy Howard is one of the experts in that area. Judy is very concerned that those kids will probably not fall into the category of "mental retardation" because they are not severe enough. She thinks that a number of them may fall into the "learning disability" category or not have problems that are readily identifiable in existing special education categories.

It is not clear whether they will have a discrepancy between their IQ and achievement. It looks to me like some of these kids will look like kids who have attention deficit disorders. In California, we do not exclude kids with emotional/behavior disorders from the LD category. So those children with parental substance abuse might fall into the learning disability category and be served there. So everybody is a little bit worried about what that portends for the diminishing resources and the increasing number of kids, especially in the LD category. The most recent figures I have seen suggest that forty percent of live births at Martin Luther King Hospital now are crack cocaine babies. The overall national figures in inner-city hospitals are ten percent of live births. So you can see the magnitude of the problem.

Question: Based on the new information you said about the finding that kids do not outgrow ADD and go through adulthood with the same symptoms, what does that mean in terms of the child who is on Ritalin who initially was told, "a couple of years ought to do it?" What is the danger of keeping the child on Ritalin for long periods of time?

Dr. Forness: It is not feared, as it once was, that Ritalin may be harmful over the long term. Earlier studies suggested that there may be diminution in growth with some of these kids, some may not gain weight nor height as rapidly as other kids who are not on Ritalin. Those studies' conclusions may have been premature. It turns out that there may not be that many adverse effects if Ritalin is used wisely, which means having a good child psychiatrist or pediatrician who knows how to use Ritalin, who knows how to use the minimal doses to get the best effect, and who gives the kids "drug vacations" over the summer, and other precautions. It also does not look like those kids are necessarily at risk for being drug-dependent or being drug abusers. It may be that the reason those kids are drug abusers is not because they are on Ritalin, but because they have attention deficit disorders. They abuse the drugs primarily because they have attention deficit disorders. They abuse the drugs primarily because they have poor judgement and impulse control problems and not because they are on Ritalin.

There are now adults who are being prescribed Ritalin, adults who have never been on Ritalin before. As a matter of fact, I myself was a control subject in an adult Ritalin trial at UCLA. There are a lot of studies going on, work with adolescents and with young adults who are being given Ritalin; and it seems to be very therapeutic. Now again, I have to hedge my bets on saying that we are still studying its use with adults, so all the data is not complete. I have to also tell you, as I said before, you have to have a good pediatrician or pediatric neurologist or child psychiatrist who knows what they are doing with the drug.

Question: You made reference to an unfortunate prognosis for adult ADD persons. Could you elaborate on that, assuming proper Ritalin control, for example, and coping strategies throughout adolescence, what is the prognosis beyond that?

Dr. Forness: For a good outcome in attention deficit disorders, it is very clear that there needs to be three components to treatment. One is medication, another is family therapy and/or parent training and the third is educational therapies and classroom modifications. These modalities of therapy seem essential for a good outcome. Even these do not always guarantee a good outcome, but on the follow-up studies, the folks who had all three different modalities combined for reasonable periods of time seemed to have a good adjustment at follow-up as adults. If you drop out any of those therapies, then the prognosis drops dramatically. If you use two of the three, then the prognosis is not good, but not as grim as nothing at all.

Generally speaking, the long-term prognosis for untreated ADD in adolescence or adulthood may be the appearance of disorders such as depression or other psychiatric disorders, underemployment, or even criminality.

Question: If treatment should continue through high school level, can you hope to have the child back into the full mainstream by late adolescence?

Dr. Forness: The issue here has to do with the variability of all these symptoms. If you remember, the diagnostic process I discussed for attention deficit disorders or conduct disorders might involve as many as fourteen symptoms. So there are all sorts of reasons to kind of hedge your bets when making predictions. The best I can tell you is that, with early and effective treatment, some cases may well involve a good outcome even well before late adolescence. Others may continue to need treatment well into adulthood.

Question: I was just wondering about treatment of ADHD within the future. I know at UCI Child Development Center, it seems like I heard of their using behavior modification, cognitive behavior, generalization maintenance, parent training. Do you see some of these components being transferred to Public Schools? I know Dr. James Swanson has just thirty percent of his kids on medication there. It just seems like the technology is out there to help kids almost normalize.

Dr. Forness: There are a number of people who are trying to initiate effective programs in the public schools. As you know, the U.S. Department of Education has given grants on attention deficit disorders and made recommendations for both regular and special education. Hopefully, one of their recommendations will be to provide specific training to teachers on the kinds of things that seem to work with these kids so that they can be initiated early on in regular classrooms and even delivered in more intensive fashion in special education classrooms.

Question: It seems like the techniques are out there in the literature. I think I have read some of these things since the late 1970's.

Dr. Forness: The techniques are in the literature, not only for this, but for several related areas in learning disabilities. There is, however, a tremendous lag between their development and their implementation into the regular classroom, and that is the problem. We currently have the will and the way, but effective ways of training teachers in knowledge and skill to deliver these services, and the funds to insure that these aspects of school reform are carried out are not always available.

Question: Do you think group therapy is more effective than individual therapy for children with learning disabilities and emotional disorders?

Dr. Forness: At particular ages, group therapy may sometimes be much more important than individual therapy, because peers may be a better source of support and insight, especially peers who have the same problem. It can sometimes be a lot more therapeutic than telling troubles to a shrink who is an adult and with whom a youngster may not necessarily feel comfortable. Group therapy and individual therapy may often be recommended in tandem, particularly when insights gained in individual therapy need to be "tried out" in the safe environment of supportive group therapy. I cannot really recommend one over the other, but I can tell you that group therapy is a really very important adjunct, particularly with kids with learning disabilities who do not often have a chance to sit down and talk about the problems they are having, not only problems with reading and writing, but about making friends, feelings of isolation or frustration, and the myriad problems associated with growing up, all of which can be complicated by having a learning disability.

REFERENCES

American Psychiatric Association (1987). *Diagnostic and Statistical Manual of Mental Disorders (Third Edition - Revised).* Washington, D.C.: Author.

Forness, S. 1988. School characteristics of children and adolescents with depression. *Monographs in Behavioral Disorders, 10,* 177-203.

Forness, S & Kavale, K. (1989). Identification and diagnostic issues in special education: A status report for child psychiatrists. *Child Psychiatry and Human Development, 19,* 279-301.

Forness, S. & Maag, J. (1991). Depression in children and adolescents: Identification, assessment and treatment. *Focus on Exceptional Children, 24,* 1-19.

Forness, S., Youpa, D., Hanna, G., Cantwell, D. & Swanson, J, (1992). Classroom instructional characteristics in attention deficit hyperactivity disorder: Comparison of pure and mixed subgroups. *Behavioral Disorders, 17,* 115-123.

Forness, S., Kavale, K. & Lopez, M. (1993). Conduct disorders in school: Special education eligibility and comorbidity. *Journal of Emotional and Behavioral Disorders, 1,* 101-108.

Kavale, K. & Forness, S. (1985). *The Science of Learning Disabilities.* Austin, TX: Pro-Ed.

U.S. Department of Education (1991). *Thirteenth Annual Report to Congress on the Implementation of the Individuals with Disabilities Act.* Washington, D.C.: U.S. Office of Special Education.

Weller, C. & Strawser, S. (1987). Adaptive behavior of subtypes of learning disabled individuals. *Journal of Special Education, 21,* 101-116.

LECTURE 5

Michael Spagna, Ph.D.
California State University, Northridge
Assistant Professor of Special Education

ALL POOR READERS ARE NOT DYSLEXIC

"Please remember not to give up on yourself."

Introduction by Richard L. Goldman

Welcome to everyone. Tonight's speaker was a teacher at Landmark West several years ago and after leaving Landmark, has considerably expanded his interests.

Dr. Michael Spagna is currently an Assistant Professor of Special Education at California State University Northridge (CSUN). He served for three years as the Coordinator of Services for Learning Disabled Students at the University of California Berkeley. He received his M. A. in Special Education at University of California Los Angeles (UCLA) and his Ph.D. from University of California Berkeley. Dr. Spagna has also studied Neuropsychology at the Medical School in the Free University in Amsterdam. He specializes in teacher training and research on learning disabilities and reading development. He has served on state advocacy committees, one of which created the publication *"I Can Learn,"* a handbook for parents, teachers, and students. His work is a unique balance between theory and practical application.

Tonight's lecture will focus on an overview of poor readers found in our schools. Our speaker will discuss the differences related to poor reader's information processing and also review applications for effective teaching. It is my privilege to introduce Dr. Michael Spagna.

Dr. Michael Spagna

I have been asked to talk tonight about poor readers and the differences that exist, if any, between what Stanovich (1990) has termed *"garden-variety"* poor readers and children who have been identified as having *"dyslexia."* I wanted to begin by asking whether or not anyone had the opportunity to read an article that appeared in the New York Times several weeks ago that reported research findings that suggest that dyslexia is not a life-long phenomenon, as previously thought. I brought the article with me tonight for those who didn't get a chance to see it.

I bring this article to your attention because it represents the type of controversy that I will address tonight; namely, does the phenomenon of *"dyslexia"* truly exist? How do individuals with dyslexia differ from non-dyslexic poor readers in terms of cognitive functioning and reading performance? And finally, if dyslexia does exist, is it curable?

In order to provide answers to these questions, it is necessary to start at the beginning: an explanation of the historical roots of the field of learning disabilities (hereafter, LD); a definition of the term *"dyslexia"*; a presentation of results from *"garden-variety"*/dyslexic poor reader comparison studies; and, a discussion of implications for teachers and parents regarding the instruction of dyslexic and non-dyslexic poor readers.

HISTORICAL ROOTS OF THE FIELD OF LEARNING DISABILITIES AND THE STUDY OF DYSLEXIA

Let us begin with a diagram depicting the origins of the LD field. As shown in Figure 1, the LD field and the study of dyslexia can be traced back to the work of two physicians named Hinshelwood and Orton. Figure 1 shows the chronological development of the study of LD and dyslexia, beginning in the 1800's.

At this point, I would like to trace the four chronological phases research in LD and dyslexia followed. Also, I want to talk about where dyslexia fits in with other types of learning disabilities such as dysgraphia, developmental aphasia, and dyscalculia.

Over the years, I have been asked the following question by teachers and parents alike: *"Is dyslexia a unique phenomenon separate from learning disabilities?"* Figure 1 should help answer this question.

Along the vertical axis of the diagram presented in Figure 1, one finds the four chronological phases of research in learning disabilities: 1) the Foundation Phase; 2) the Transition Phase; 3) the Integration Phase; and, 4) the Contemporary Phase. These phases took place from the 1800's through the 1930's, the 1930's through the 1960's, the 1960's through the 1980's, and the 1980's through the present, respectively. Along the horizontal axis, one notices different types of learning disabilities: Disorders of Spoken Language; Disorders of Written Language; and, Disorders of Perceptual and Motor Processes.

It is important to point out that one type of learning disability is conspicuously absent from Figure 1: Disorders of Mathematical Reasoning. Some of you might be familiar with the term *"dyscalculia,"* which is often associated with this type of learning disability. Disorders of mathematical reasoning are distinct from problems with reading or writing that affect math performance, and have just recently been studied closely.

The area that I am going to focus on tonight is located in the middle section of Figure 1: Disorders of Written Language. This type of learning disability is referred to today as the phenomenon of dyslexia. In the early 1900's, Hinshelwood and Orton were among the first to document children coming into their clinic who were experiencing extreme difficulty reading. These patients could not read fluently in spite of receiving normal instruction, coming from an average socio-cultural background, and

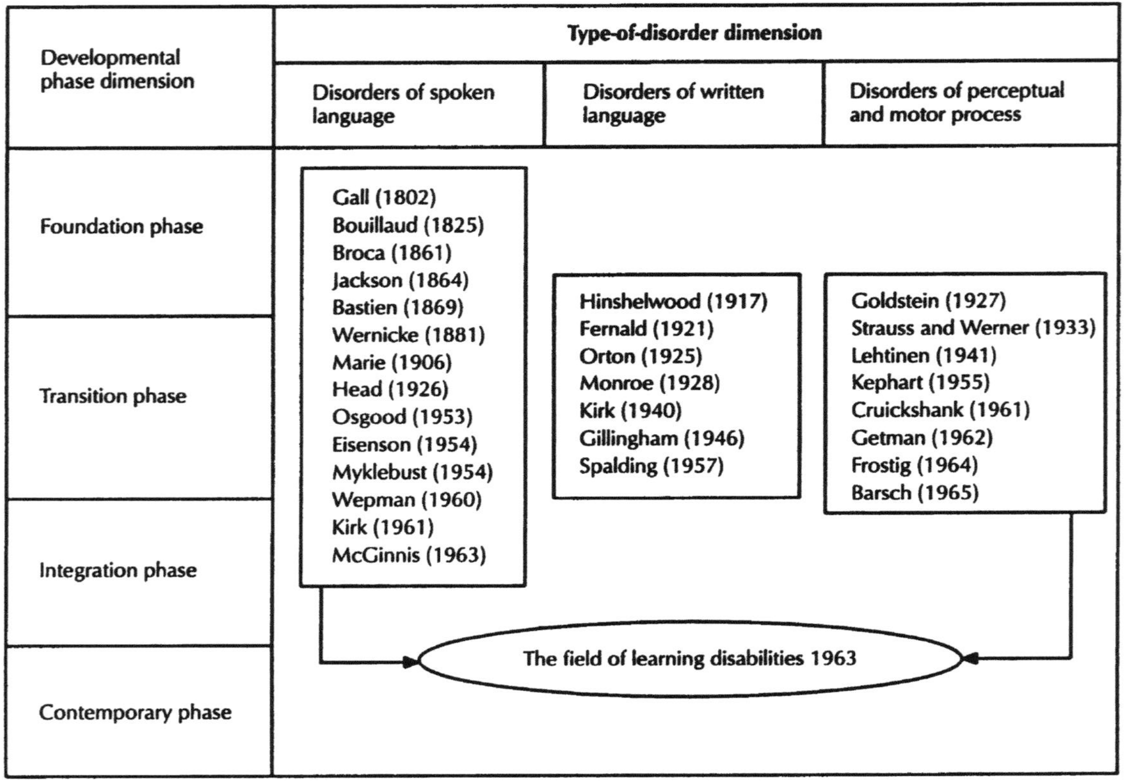

Figure 1. The four chronological phases of the study of dyslexia and the field of learning disabilities (adapted from Wiederholt, 1974, p. 105).

possessing average or above average cognitive functioning. At this time, following the example of Hinshelwood and Orton, there manuscript page were numerous case studies documenting the reading behaviors exhibited by children.

As I discuss the history of dyslexia, it is very important to emphasize the value of this initial case study research. As you will see later, researchers soon got away from these types of qualitative studies, instead opting for larger, seemingly more glamorous, quantitative projects. This initial work was invaluable since it was studied what children were doing while engaged in reading. Only by observing dyslexic children as they read can one determine whether their reading skills are qualitatively different from other types of readers.

During the Foundation Phase, Hinshelwood (1917) was among the first to publish observations, describing a thirteen year old male who was able to learn auditorially but could not remember visual symbols. Hinshelwood speculated that this *"word blindness"* was due to a neurological defect in the angular gyrus of the brain. Orton (1928) was famous for coining the term, *"strephosymbolia,"* which directly translates to *"twisted symbols."* Like Hinshelwood, Orton reported cases of children who had *"word blindness,"* able to learn oral language but unable to decode words. Orton theorized that this inability to read was not due to a brain defect but attributable to a lack of cerebral dominance.

During the Foundation Phase, and into the early 1930's, this phenomenon of dyslexia was primarily studied from a medical perspective; and, as a result, was generally believed to be a biological phenomenon, caused by neurological dysfunction. From the 1930's onward, however, professionals outside the field of medicine, primarily in the disciplines of psychology and education, became involved with the study of dyslexia. During this Transition Phase, psychologists such as Mable Fernald (1943) gained notoriety by studying the psychological characteristics of children who were having difficulties reading. In fact, Fernald, through her work at the University of California at Los Angeles (UCLA), was among the first to propose that dyslexia was the result of specific psychological abilities/processes which were dysfunctional—not operating at a level commensurate to the level of other psychological processes. Although her theory was proposed over seventy years ago, Fernald's notion of faulty information processing is in accordance with modern theories concerning dyslexia. Through work at her UCLA clinic, she developed a multi-sensory approach, called the *"VAKT"* approach (*Visual, Auditory, Kinesthetic, and Tactile*), to remediate reading and spelling difficulties.

Fernald's VAKT method was ultimately designed to help youngsters compensate for an informational processing deficit. If you haven't had the chance to read her book, *Remedial Techniques in Basic School Objectives* (1943), I highly recommend it.

Turning now to the Integration Phase, the term *"learning disabilities"* was first introduced by Samuel Kirk in 1963 to what was to become the Association of Children with Learning Disabilities (ACLD). At this time, all of a sudden, case studies on dyslexia disappeared from the research landscape. With the introduction of the term *"learning disabilities"* in 1963, federal legislation soon followed in 1969 (Public Law 91-230) and 1975 (Public Law 94-142). With federal regulations used to determine LD eligibility on the books, a rapid expansion of school programs designed to educate children identified as having dyslexia was experienced in the United States. For the first time, learning disabilities (and dyslexia) was recognized as a disability category, deserving of special education and related services.

As mentioned earlier, the first federal law was enacted in 1969, the *Children with Specific Learning Disabilities Act*, otherwise known as Public Law 91-230. When P.L. 91-230 was enacted, large numbers of youngsters were identified as having learning disabilities, and pertinent educational services were allocated. This legislation proved important from an educational standpoint; children with dyslexia were finally given access to necessary instructional interventions.

Finally, in the Contemporary Phase (1980's - present), the LD field has continued to expand with services being extended to dyslexic adults, mainstreaming and collaboration becoming the educational method of choice, and the development of LD and dyslexia organizations and advocacy groups.

THE IMPORTANCE OF CASE STUDY RESEARCH

I have started out this session tonight with a discussion of the historical roots of the LD field and dyslexia for a reason: To point out that, with the expansion of school programs during the Integration Phase, case study research was largely abandoned, remaining unfinished. Without such qualitative research, the description of characteristics of dyslexic children was left incomplete. In response to such abandonment of qualitative research, one researcher, Sabatino (1981), said that the field of learning disabilities had become a field of "service delivery in search of theory."

Actually, today, we see a re-emergence of case study research, with professionals returning to unfinished business, explaining how youngsters with dyslexia read, and whether they read differently from other poor readers. My own research follows in this line of inquiry, studying the differences between children who are dyslexic and children who are just *"garden-variety"* poor readers.

Before I leave this discussion of the importance of case study research, let me mention the influential work of Elena Boder (1973) which remains largely overlooked today. Boder is a child psychiatrist here in Los Angeles who looked at the reading performance of children who had been identified as dyslexic, and purportedly discovered three types of dyslexia. I shall return to Boder's three types of dyslexia later when I talk about implications for instruction, because I believe the investigation of reading differences provide insight into what goes on with youngsters who are poor whole-word readers, versus children who are poor decoders.

BODER AND DYSLEXIA SUB-TYPES

Boder identified the following dyslexia sub-types: 1) *"Dysphonetic"* dyslexia; 2) *"Dyseidetic"* dyslexia; and, 3) a *"Mixed"* group. "dysphonetic" dyslexia (67% of total dyslexics studied) were described as those children who exhibited a lack of *"phonological awareness,"* with the letters we see on a page, *"graphemes."* Other researchers have referred to these readers as having *"deep dyslexia"* (Coltheart, Patterson, & Marshall, 1980). Let me give you an example of an error this type of reader might make: Seeing the word *"cape,"* and pronouncing it as *"cap,"* failing to recognize the silent *"e."* Boder explained that these readers usually substitute a whole word that they are familiar with for an unfamiliar word that is difficult to decode. A second type of dyslexic reader Boder reported, though not in as great numbers, was termed a *"Dyseidetic"* reader (10% of total dyslexics studied). Later, researchers referred to this group of readers as exhibiting *"surface dyslexia"* (Coltheart et al., 1980). These youngsters were described as being unable to read sight words effectively. However, they could pronounce and read words that required *"phonological awareness."* For example, these children, when exposed to the word *"steak"* (pronounced correctly as *"stak,"* would pronounce it as *"stek."* Boder provided the following explanation of this phenomenon: *"Dyseidetic"* dyslexics are aware of the phonological rule that *"ea"* can be pronounced as *"e"* in certain words and, therefore, have applied the rule in this instance. In other words, these readers have difficulty with sight words, which require previous exposure and visual memorization.

Finally, the third type of dyslexia described by Boder was the *"Mixed"* group (23% of total dyslexics studied). These children not only exhibited a poor sight vocabulary, but also experienced problems sounding out words. As a result of the severity of reading problems in these youngsters, Boder also offered the worst educational prognosis for this group: *"Mixed"* group dyslexic children could not rely on sight-word or phonic skills to compensate for word recognition/decoding weaknesses.

Before I leave Boder's work, which I think is very important, I want to point out some additional findings that were reported after Boder's initial research. This research posed the question: *"How would children from the three dyslexia groups read irregular words?"* An example of an irregular word would be the word *"laugh"* which is a sight word that cannot be decoded using *"regular"* rules of phonics. If one tries to sound out *"laugh"* using only phonological skills, one might pronounce it as *"lag,"* interpreting the graphemes *"gh"* as making the *"g"* sound/phoneme. According to Boder's work, what might one expect regarding the performance of *"Dysphonetic"* and *"Dyseidetic"* dyslexic readers when exposed to sight words such as *"laugh?"* In other words, which type of dyslexic reader would be expected to have more difficulty reading sight words?

If you said the *"Dyseidetic"* group, you would be correct since these youngsters would, most likely, experience difficulty remembering the pronunciation of the word as a whole unit. Instead, *Dyseidetic"* readers attempt to sound out unfamiliar words—relying on phonics, which produce incorrect pronunciations such as *"lag."* The *"Dysphonetic"* group, on the other hand, could probably read *"laugh"* correctly, especially if they had been previously exposed to it. Remember that in the *"Dysphonetic"* group, the knowledge of sight words is stronger than the ability to sound out words.

Let me give you another example which I think will sharpen the discrimination between *"Dysphonetic"* and *"Dyseidetic"* dyslexics. *"Pseudo-words"* or *"nonsense"* words are words that are fairly regular, phonetically speaking. Usually, a *"real"* word is changed slightly to create a new *"nonsense"* word; for instance, if I take the word *"jump,"* remove the first letter *"j"* and replace it with a *","* I have created the *"nonsense"* word *"tump."* If one were to construct a list of these *"nonsense"* words and present it to the three groups of dyslexic youngsters described by Boder, which group would you expect to have more difficulty reading these *"nonsense"* words? The *"Dysphonetic"* or the *"Dyseidetic"* group?

Using Boder's findings, and according to her theory, *"Dysphonetic"* dyslexics should have more difficulty reading *"nonsense"* words, since they lack strong phonological skills.

Boder's work should be acknowledged as a significant contribution to the study of dyslexia since it provided considerable insight into how children read. To finish with Figure 1, research conducted during the Contemporary Phase will most likely pose *"why"* questions instead of *"how"* questions, not just investigating *"how"* youngsters read sight words and *"nonsense"* words ala Boder, but also examining *"why"* children read words in the manner that they do. Obviously, some children won't be able to verbalize what they're doing when reading words; but others can. And I shall describe later, in my own research, what children have told me when asked *"why"* they pronounce words in a certain manner.

INCIDENCE AND A DEFINITION OF DYSLEXIA

Before continuing with my talk tonight, let me define the term *"dyslexia"* for you according to the World Federation of Neurology (cited in Shaywitz, Escobar, Shaywitz, Fletcher, & Makuch, 1992): *"a disorder manifested by failure to attain the language skills of reading, writing and spelling, despite conventional instruction, adequate intelligence and socio-cultural opportunity."*

Now, to turn to present day information about dyslexia, I would like to discuss the incidence of children with dyslexia in California and the United States. Due to Public Law 91-230, as well as ensuing legislation (P.L. 94-142 and Section 504 of the Rehabilitation Act of 1973), there has been an explosion in the number of youngsters identified as having disabilities. Currently, in the state of California, approximately 5% of the school-age population has been identified as having LD (USOSERS, 1991). This represents approximately a 140% increase since 1975 in the number of children identified as LD. Faced with this information, however, one must ask: *"Of those identified as LD, how many youngsters are poor readers, and possibly dyslexic?"* In studies of school-age children, between 2% and 6% are usually found to have dyslexia (Yule & Rutter, 1975). Applying these findings to the figures reported earlier, it could be argued that anywhere between 40% and 100% of the LD population has dyslexia.

Also, when confronted with such a staggering increase in the number of LD-identified children, one must also ask: *"Are all those identified as LD, truly LD?"* One study that addresses this issue was conducted by Shepard, Smith & Vojir (1983) in Colorado. Shepard et al.'s findings appear in Figure 2. Shepard et al. attempted to determine whether or not children who were being labelled learning disabled truly were, using psychometric data collected at various school sites. You'll notice in Figure 2, under the learning disabilities category, that Shepard et al. found less than 50% of the population studied, using Colorado identification criteria, were truly learning disabled. Many children identified as LD should have properly been diagnosed in one of the other categories listed in Figure 2: Other Handicaps; Other Learning Problems; or, Other.

I don't have time tonight to talk about the reasons for this, but, suffice to say that teachers and administrators are faced with a very hard decision: *"I have these students who do not meet the criteria exactly but they're not making it in my class/school." "What am I going to do?"* Simply put, many youngsters are identified as LD because this is an avenue to provide educational assistance; without the label, teachers and administrators are faced with the very real alternative—early drop-out. Faced with such a decision, Shepard's findings appear less surprising. However, with the possibility for a greater number of *"false-positive"* LD identifications—children who are not LD being identified as LD—educators and researchers are, in turn, faced with an equally difficult proposition: studying a heterogeneous population which might contain less than 50% *"true"* LD children. Obviously, the implications for research and education are enormous.

Now, let's talk about incidence studies of dyslexia. Reported in the literature, there was a landmark study, conducted in England by Rutter and Yule (1975), known as the *"Isle of Wight Study."* These researchers tested approximately thirty-four hundred 9, 10 and 11-year-olds to study their reading skills, cognitive functioning, and so forth. Rutter and Yule reported that certain youngsters exhibited high levels of general cognitive functioning but comparatively lower levels of reading skill. These children as children were referred to by Rutter and Yule as having *"Specific Reading Retardation"* (hereafter, SRR). Moreover, Rutter and Yule differentiated these youngsters from children who were not only poor readers but also demonstrated lower levels of cognitive functioning; this second group of poor readers was identified as having *"General Reading Backwardness"* (hereafter, GRB).

This 1975 study is important because Rutter and Yule were among the first researchers to discriminate between two groups of poor readers: SRR children with high-level cognitive functioning who were poor readers (we normally consider this group to be dyslexic poor readers), and GRB youngsters who had low cognitive functioning across the board, in-

	Percentage of LD Cases	Standard Error
Other Handicaps		
Educable Mentally Retarded	2.6	+0.6
Emotionally Disturbed	7.5	+1.0
Hearing Handicapped	0.2	+0.2
(Total)	**10.3**	
Learning Disabilities		
Significant Ability/Achievement Discrepancy	20.5	+2.0
High Quality Processing Deficit	4.7	+0.8
Brain Injured	0.6	+0.3
Hyperactive	2.0	+0.6
Weak Significant Discrepancy and Verbal/ Performance Discrepancy	3.6	+0.6
Weak Significant Discrepancy and Medium Quality Processing Discrepancy	1.1	+0.4
Medium Quality Processing Discrepancy and Verbal/Performance Discrepancy	6.6	+1.2
Medium Quality Processing Discrepancy	3.5	+0.8
(Total)	**42.6**	
Other Learning Problems		
Language Interference	6.6	+1.0
Slow Learners	11.4	+1.4
Environmental Causes	2.2	+0.6
Below Grade Level Achievement	6.1	+1.0
Minor Behavioral Problems	3.7	+0.8
(Total)	**30.0**	
Other		
Poor Assessment (no IQ and no Ach. tests)	6.4	+1.3
Miscellaneous (including normal)	10.6	+1.3
(Total)	**17.0**	

Figure 2. Identification of learning-problem subgroups in the Colorado Study (Shepard, Smith, & Vojir, 1983).

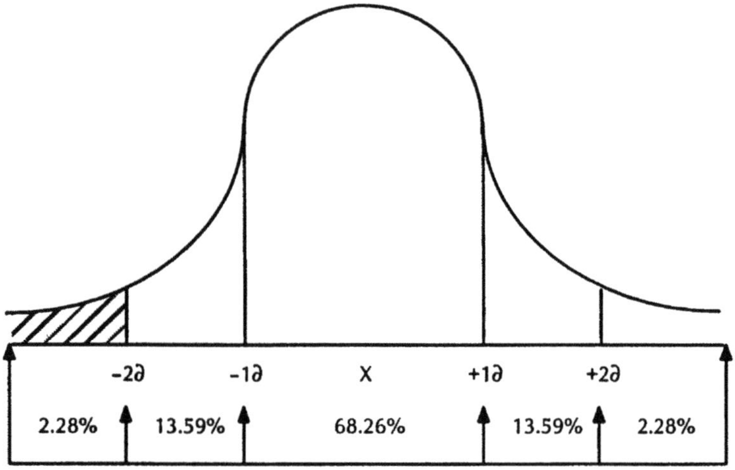

Figure 3. Normal distribution of reading ability among the school-age poulation.

cluding lower reading ability (we normally consider this group to be *"garden-variety"* poor readers). Of course, the insinuation with SRR children is that something's gone wrong since a relatively high level of cognitive functioning exists. Why aren't SRR children reading up to their ability? It is not surprising that the child with low levels of cognitive functioning also exhibits low level reading skills.

I bring this study to your attention for the following reason: When Rutter and Yule presented their findings, quite a bit of controversy was generated. It had previously been thought that only one group—not two—of poor readers existed, the GRB group.

To further understand the importance of Rutter and Yule's study, I shall describe their methodology in more detail. Rutter and Yule examined all of the Isle of Wight youngsters who had a certain IQ level and looked to see how they were distributed in terms of their reading ability. Looking at a normal curve, such as the one presented in Figure 3, one would expect all of the children having the same level of intelligence to exhibit a reading ability distribution similar to a normal curve. It is important to notice that the shaded area appearing in Figure 3 represents a group of children who were found to have reading skills *"two standard deviations below the mean."* In other words, these youngsters have significantly below average reading skills. If one were to think that the reading skills of children

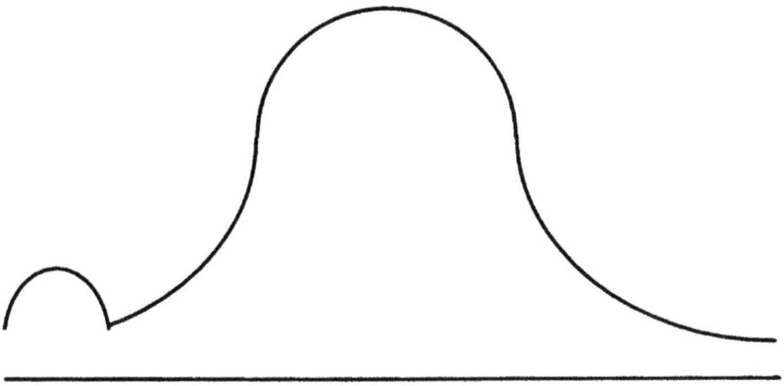

Figure 4. Bi-modal distribution of reading ability reported by Rutter and Yule.

of similar IQ were distributed along a normal curve, one should also expect that only 2.28% of this same population of youngsters would have severe reading problems *"two standard deviations below the mean."*

When Rutter and Yule conducted their research, they discovered a greater incidence of children with severe reading problems than would be expected using hypothetical distribution of reading skill depicted in Figure 3. Indeed, Rutter and Yule found that approximately 9% of the Isle of Wight population exhibited severe reading problems—not the 2.28$ expected. Instead of a normal distribution, Rutter and Yule reported that reading skills could better be described as comprising a bimodal distribution, with a smaller crest appearing towards the bottom of the curve. This bimodal distribution is depicted in Figure 4.

The conclusion offered by Rutter and Yule—namely, that reading performance was not evenly distributed among youngsters of similar IQ—was revolutionary at the time. Pointing to these findings, researchers all over the world believed that the *"Holy Grail"* of dyslexia had finally been discovered, proving that a larger group of children—comprised partially of dyslexic individuals—were experiencing reading problems than would be expected. In several professional circles, this finding validated the long-held belief that a *"true"* phenomenon of dyslexia existed. Furthermore, Rutter and Yule's work was interpreted to indicate that these dyslexic poor readers were distinct from general poor readers.

Several researchers have disputed Rutter and Yule's findings, however (most notably, Rodgers, 1983). If we return to Figure 4 for a moment, I will explain why Rutter and Yule's findings and, hence, their conclusions have purportedly failed to hold up under scrutiny.

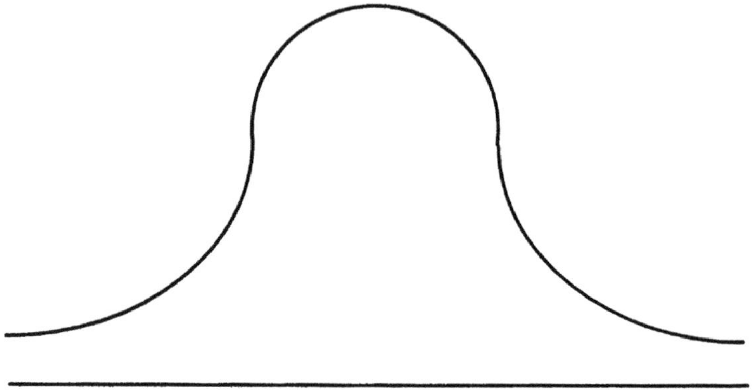

Figure 5. Normal distribution of reading ability after removing ceiling effects (Rodgers, 1983).

You will notice that the curve presented in Figure 4, when compared to the curve shown in Figure 3, is skewed to the right. One possible reason for the skew would be that the test was too easy; in other words, the majority of children taking the test scored well on the assessment instrument. Educational measurement experts refer to this phenomenon as a *"ceiling effect,"* with many subjects scoring manuscript toward the upper limit of the test. When *"ceiling effects"* are present, it can be inferred that the measurement device used fails to maximally discriminate between the performance of subjects.

Applying an understanding of this phenomenon to Rutter and Yule's findings, one can justifiably offer a competing interpretation of the significance of a bimodal distribution: Rutter and Yule's study was methodologically flawed; they failed to use a more appropriate measurement device, one that guards against *"ceiling effects."* Rodgers (1983) went back and reanalyzed the finding of Rutter and Yule, controlling for *"ceiling effects"* on the reading skill measurement device. A new distribution of reading scores was generated by Rodgers which is presented in Figure 5. As you can see in Figure 5, once *"ceiling effects"* are removed from the data, a normal, not a skewed curve results. This finding directly contradicts those presented by Rutter and Yule, in effect, pointing away from the conclusion that an unexpected dyslexic group of poor readers exists. So, again, one witnesses another controversy being created, with people doubting the existence of dyslexia.

According to Rodgers' work, one could conclude that there is no such phenomenon as dyslexia, especially since Figure 5 doesn't portray an unexpected abnormal group of poor readers at the bottom of the distribution.

As a result of Rodgers' study, the findings reported by Rutter and Yule are pretty much discounted by professionals in the field. This is quite unfortunate, since Rutter and Yule gathered additional information in their longitudinal study that is valuable.

After an initial examination of the cognitive and reading performance of the "Isle of Wight" population, Rutter and Yule re-administered all tests five years later. Based on this longitudinal information, Rutter and Yule offered an educational prognosis for the SRR and GRB groups: five years later, SRR youngsters continued to experience difficulty with their reading (even lower performance than their GRB counterparts), but also exhibited mathematics skills approaching grade level expectations; GRB children, on the other hand, scored well below grade level on the follow-up tests of reading and mathematics.

These additional follow-up findings prove important to educators, and, to a certain extent, are more significant than the previously discussed incidence findings; comparing the performance of SRR and GRB children, the two groups have different educational prognoses, with SRR youngsters requiring more intensive instruction in reading although remaining at grade-level in other academic areas (mathematics).

"PHONECIAN" AND "CHINESE" READERS: A TWO-DIMENSIONAL CONTINUUM OF READING SKILLS

For the last seventeen years, thanks to the work of Rutter and Yule, researchers investigated the reading performance of children, searching for ways to discriminate between types of readers. Baron and Strawson (1976) were among the first to describe ""*normal*" readers as belonging to one of two groups: "*Phonecians*" and "*Chinese*" readers. To be more specific, Baron and Strawson described Phonecian readers as being better alphabetic readers, using a "*sounding-out*" approach to decode words; hence, this group of readers could be characterized as having a relatively high level of phonological awareness. Chinese readers, on the other hand, were characterized as being good sight-readers, recognizing words visually as single units having a unique pronunciation. Another way of describing these two groups of readers would be to characterize "*Phonecian*" readers as relying on an "*orthographic system*" to decode words, using knowledge of phoneme/grapheme correspondences while reading; "*Chi-*

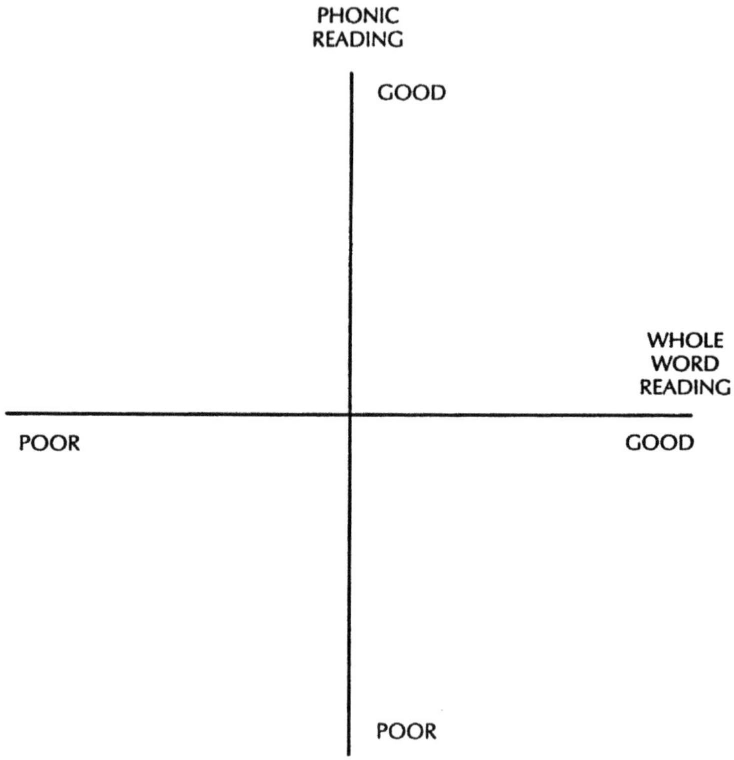

Figure 6. A two-dimensional continuum of reading skills (Ellis, 1985).

nese" readers, in comparison, would be described as relying on a *"logogaphic system"* to read, using visual-memory to access pronunciations of words as whole units. Figure 6 (Ellis, 1985) illustrates reading skills using a two-dimensional chart; the vertical axis represents *"phonic reading"* skill (sounding out approach) and the horizontal axis represents *"whole word reading"* skill (sight-word approach). The top of the vertical axis represents good *"Phonic reading"* skill; the bottom of the vertical axis indicates poor *"phonic reading"* skill. To complete the diagram, the left side of the horizontal axis represents poor *"whole word reading"* skill, while the right side of the horizontal axis indicates good *"whole word reading"* skill. Is there anyone in the audience tonight who can guess where Baron and Strawson's *"Chinese"* readers would exist on this diagram? Looking for a moment at Figure 7, you will notice that this group can be found in the

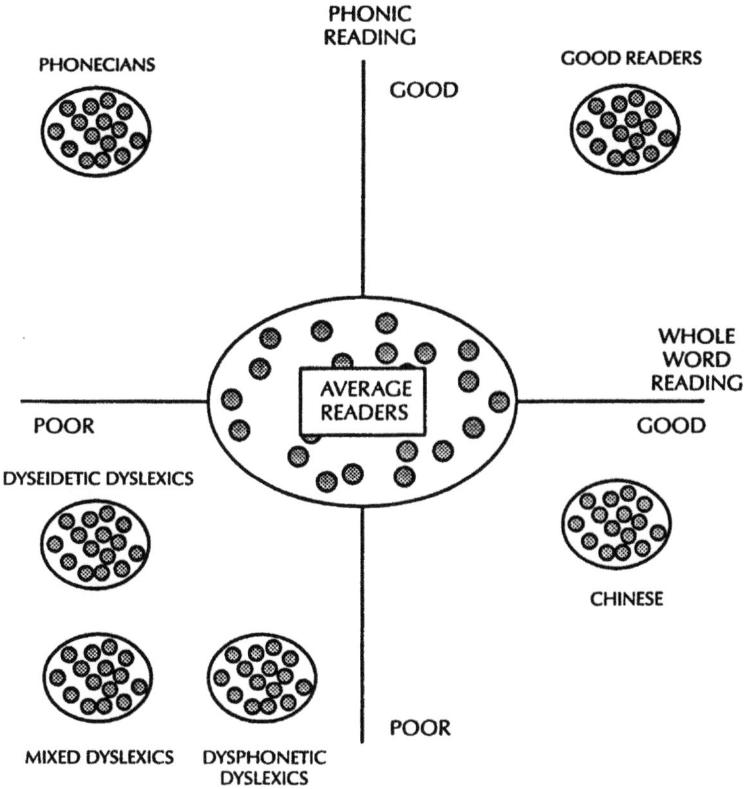

Figure 7. Hypothetical distribution of average, good, Chinese, Phoecian, and dyslexic readers; homogeneous continuum (Ellis, 1985).

middle-left part of the graph. The *"Chinese"* readers belong here since they exhibit good whole word reading skill, but only average phonic reading skill.

Returning again to Figure 7, where do you think the *"Phonecian"* reading group belongs? Remember that this reader group has good phonic reading skills but not necessarily good whole word reading skills. According to Baron and Strawson's work, this reader group would be best placed in the upper-middle to upper-left portion of the diagram.

Now, finally, where would you put the students who are dyslexic readers, *"Dysphonetic," "Dyseidetic"* or *"Mixed"* groups. As Figure 7 depicts,

"Dysphonetic" readers—who have poor phonic reading skills but average whole word reading skills—are placed at the lower-middle part of the diagram. *"Dyseidetic"* dyslexics, on the other hand, appear at the middle-left section of the figure since these readers exhibit poor whole word reading skills and average phonic reading skills. The third group of dyslexic readers, the *"Mixed"* group, appears at the lower-left part of the graph; this group exhibits poor phonic reading skills as well as poor whole word reading skills.

To complete the diagram presented in Figure 7, notice that *"Good"* readers appear in the upper-right section of the graph since they have good phonic and whole word reading skills. It is important to point out here that terms like *"Dysphonetic"* dyslexics, *"Dyseidetic"* dyslexics, *"Chinese"* readers, and *"Phonecian"* readers, one might be left with the impression that each term identifies separate, distinct groups of readers, with little or no overlap between members of the different groups. Current thinking (see Ellis, 1985) dictates that good and poor readers do not fall into separate, homogeneous groups, such as those depicted in Figure 7. Instead, reader groups are distributed in a fashion similar to that depicted in Figure 8; children fall along a heterogeneous continuum, with quite a lot of overlap between different skill levels.

Figure 8 clearly illustrates how children differ from one another in their reading skills, using the two axes of phonic reading and whole word reading to create a reading-skill continuum. Experts in the field, for the most part, agree that all readers belong in one of the Figure 8 reading skills quadrants. Some youngsters are not going to be great sight-word readers; others are going to experience difficulty sounding out words. But, by using both of these dimensions, you can kind of get a picture of where children might fall along this continuum.

DYSLEXIC VS. "GARDEN-VARIETY" POOR READER COMPARISON STUDIES

Several studies have been conducted, examining the reading performance of children who would appear in the lower-left hand section of Figure 8, but not belonging to the dyslexia categories. These non-dyslexic youngsters have been identified as *"Garden-variety"* poor readers (Stanovich, 1990). Similar to the youngsters identified by Rutter and Yule as belonging to the *"General Reading Backwardness"* (GRB) group, the *"Garden-variety"* group exhibits below average cognitive functioning and below average reading skills. For the most part, studies examining the performance of

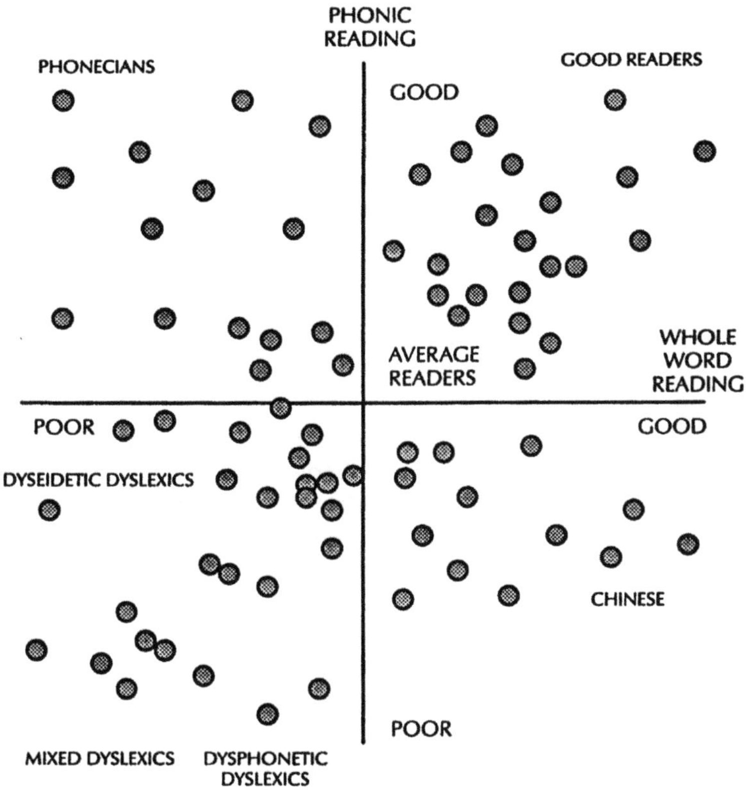

Figure 8. Hypothetical distribution of average, good, Chinese, Phoecian, and dyslexic readers; heterogeneous continuum (Ellis, 1985).

this *"Garden-variety"* group have compared these youngsters to children who have been identified as having some form of dyslexia or *"specific reading retardation"* (children who have average or above average cognitive functioning but below average reading skills). Two such studies, one conducted at the University of Kansas (Warner, Schumaker, Alley, & Deshler, 1980) and one at the University of Minnesota (Ysseldyke, Algozzine, Shinn, & McGue, (1982), compared the cognitive functioning and reading performance of dyslexic poor readers and *"Garden-variety"* poor readers.

Considering the information I have presented so far about youngsters who would be classified as either dyslexic or *"Garden-variety"* poor readers, what would you speculate were the findings of these studies? In other words, what do you think the two research groups found when comparing the cognitive functioning and reading skills of the two groups?

Based on the identification criteria already set out, one would expect both groups to exhibit below average reading skills; one would also expect the dyslexic group to have higher levels of cognitive functioning when compared to the *"Garden-variety"* group. When reviewing the results obtained from the Kansas and Minnesota studies, however, one is faced with quite unexpected findings.

First of all, both research groups failed to report significant cognitive functioning differences between the two groups. However, the dyslexic group scored at a lower level than the *"Garden-variety"* group on reading tests. In other words, one might interpret these findings to indicate that the children who were identified as having dyslexia were, simply put, the poorest of the poor readers. Do these findings make sense considering our seemingly incorrect predictions? Why were the results opposite from what we predicted would happen? A closer look at the University of Kansas and the University of Minnesota studies reveals the answers to these questions.

Neither study selected youngsters using criteria similar to that described earlier in Rutter and Yule's work; namely, discriminating between dyslexic and *"Garden-variety"* poor readers using differences in cognitive functioning to separate the two groups. Instead, the University of Kansas and University of Minnesota studies relied upon school-based identification criteria to separate dyslexic and non-dyslexic poor readers. So, a better interpretation of the findings reported in the Kansas and Minnesota studies would be the following: The school sites sampled identified children experiencing the most difficulty reading as having dyslexia, without regard to differing levels of cognitive functioning. It is important to point out that schools, faced with dwindling financial resources, might identify those children who require the greatest educational intervention as having *"dyslexia."* Such a decision would obviously by influenced by economic concerns first and foremost.

Any of you who have ever worked with dyslexic youngsters, however, have come to find that differences do exist between these children and *"Garden-variety"* poor readers. Simply describing children with dyslexia as the poorest of the poor readers is an inaccurate statements. If intellectual differences did not exist, then one would expect that a dyslexic youngster would exhibit cognitive functioning commensurate with his/her level of reading skill—below average in both areas.

My experience with dyslexic children, however, fails to support this predicted level of performance: Dyslexic readers have a rich knowledge of the world, what's going on around them; they can extract information and exhibit average to above average levels of cognitive functioning. Instead of demonstrating generalized deficits in such cognitive areas as memory, linguistic comprehension, etc., dyslexic individuals experience a specific type of learning problem either in their understanding of phonics or in their ability to remember whole word units. *"Garden-variety"* poor readers, in comparison, do experience generalized cognitive deficits that not only affect memory and linguistic skills, but which also affect reading skills.

Differences in cognitive functioning between dyslexic and *"Garden-variety"* poor readers will directly influence the effectiveness of educational interventions: A student with low cognitive functioning and low reading skills will require educational experiences that will build his/her knowledge of the world; with a dyslexic student, however, it is more important to tap into existing knowledge of the world, using his/her experiential background to make reading more meaningful. This is not to say that *"garden-variety"* poor readers do not possess significant knowledge of the world around them; they do. But it is also important to emphasize that dyslexic youngsters, as described earlier, are more likely to have acquired even more knowledge than their *"Garden-variety"* counterparts. The secret to teaching reading to both groups is to unlock the seemingly hidden background knowledge that these youngsters possess and to demonstrate how new knowledge, gained through reading, can be linked to what they already know.

A "Simple" Model of Reading

In order to understand how dyslexic and non-dyslexic poor readers might differ in their reading skills, I think it's helpful to show you an equation, proposed by Gough and Tunmer (1986), that describes the relation between decoding words—a source of extreme frustration to dyslexic readers—and reading. Essentially, Gough and Tunmer believed that reading comprehension (R) is the product of decoding (D) and linguistic comprehension (C); hence, the equation: $R = D \times C$. by linguistic comprehension, Gough and Tunmer refer to the individual's understanding of language, and the world around him/her.

Using this reading equation, with each variable ranging from 0 (lowest skill level) to 1 (highest skill level), Gough and Tunmer proceed to describe the different types of readers encountered among the school-age population: *"Normal"* readers; *"Dyslexic"* readers; *"Hyperlexic"* readers; and *"Garden-variety"* poor readers. A *"Normal"* reader would be characterized as having a high level of decoding (1.0) and linguistic comprehension (1.0). The predicted reading comprehension of this *"Normal"* reader would be equal to the product of these two components; hence, 1.0 (decoding) × 1.0 (linguistic comprehension) = 1.0 (reading comprehension).

In comparison, a Dyslexic reader would be expected to exhibit poor decoding skill; let's say, for illustrative purposes, that a dyslexic reader has a decoding level of 0.40, less than half of that exhibited by normal readers. The dyslexic person's linguistic comprehension, his/her knowledge of the world and what he/she can pick up in conversation however, compared to normal readers, is nearly identical: linguistic comprehension equals 0.9. Multiplying those two variables, one arrives at a substandard reading comprehension level: 0.36—significantly below a perfect score of 1.0

"Garden-variety" poor readers, in comparison to dyslexic readers, are characterized by Gough and Tunmer as having relatively higher levels of decoding skill—say 0.60—and lower levels of linguistic comprehension—say, 0.60. According to the reading equation, one would predict that the reading comprehension of *"Garden-variety"* poor readers—0.36—would be equal to that of Dyslexic readers.

Finally, Gough and Tunmer describe a fourth type of reader, the Hyperlexic reader. This type of reader is characterized as having relatively high levels of decoding skill (e.g., Decoding equal to 0.90) and low levels of linguistic comprehension (e.g., Linguistic comprehension equal to 0.40). Usually, the kind of individuals we see who have hyperlexia are children having moderate to severe brain damage: children who have had some kind of neurological insult or people who have had strokes, maintaining some of their decoding ability but losing general language skills. Using Gough and Tunmer's equation, one would predict that a Hyperlexic reader would exhibit a reading comprehension level of 0.36, identical to the levels found for *"Dyslexic"* and *"Garden-variety"* poor readers.

Considering the reading comprehension levels of the three groups of poor readers—all at 0.36, respectively—it is necessary to discuss educational implications for these readers. Think about it—is it easier to build up a child's understanding of the world, or is it easier to teach decoding skills? If it were possible to remediate the decoding skill deficit of all three poor reader types, changing decoding values to 1.0, which group would exhibit the highest level of reading comprehension?

Turning back one final time to the reading equation proposed by Gough and Tunmer, we can see the new reading comprehension levels of the three poor reader groups if the decoding variable is set at a perfect value of 1.0 for each group; previous reading comprehension levels will be listed in parentheses.

"Dyslexic Readers":	1.0 (D) × 0.9 (C) = 0.9 (R)	[0.36]
"Garden-variety" Poor Readers:	1.0 (D) × 0.6 (C) = 0.6 (R)	[0.36]
"Hyperlexic" Readers:	1.0 (D) × 0.4 (C) = 0.4 (R)	[0.36]

It becomes quite obvious that, if the equation proposed by Gough and Tunmer proves valid, Dyslexic readers have the best prognosis for educational success, as far as improved reading is concerned once their decoding deficits are remediated.

Decoding Strategies of "Dyslexic" and "Garden-Variety" Poor Readers

So far I have been talking about theoretical models and "simple" equations to describe different types of readers. It all seems logical on paper and in theory. But, how does one test these theories to ascertain their veracity? If the theoretical reading continuum, presented in Figure 8, really holds up, one should be able to interview and test youngsters to investigate whether or not qualitative differences do exist between *"dyslexic"* and *"garden-variety"* poor readers. Sticking to the theory, *"garden–variety"* poor readers should have poor decoding skills as well as low levels of linguistic comprehension; *"dyslexic"* readers, in comparison, should exhibit similar decoding deficits, but should also have higher levels of linguistic comprehension. In other words, *"dyslexic"* youngsters should be poor decoders and have a relatively good understanding of the real world.

In my research, I have investigated the decoding skills and linguistic comprehension of *"dyslexic"* and *"garden-variety"* poor readers. There were two phase to a project I began about a year and a half ago. The first phase began with a meeting I was fortunate to have with Richard Woodcock, who created the Woodcock Reading Mastery Test. Woodcock afforded me access to his database of 6,000 people used to standardize the Woodcock-Johnson Psycho-Educational Battery-Revised (1989). In order to compare the reading performance of school-age youngsters, I separated children into three groups: good readers, *"dyslexic"* poor readers, and *"garden-variety"* poor readers. The research involved an examination of the performance of *"dyslexic"* and *"garden-variety"* poor readers on one of the subtests of the Woodcock-Johnson Psycho-Educational Battery, the Word Attack subtest (a test of *"nonsense"* words). It is believed that the best measure of phonological awareness involves *"nonsense"* or pseudo-words; since these pseudo-words are unfamiliar to the reader, he/she cannot rely on a sight-word approach and must decode them using phonological knowledge. Boder's research (1973) has shown that youngsters who can be classified as having *"Dysphonetic dyslexia"* will encounter difficulty reading such *"nonsense"* words aloud.

The Woodcock-Johnson Psycho-Educational Battery includes a list of pseudo-words in the Word Attack subtest. Figure 9 lists all of the *"nonsense"* words included in the Word Attack subtest; notice that the pseudo-words become increasingly difficult beginning with a simple consonant-vowel-consonant (CVC) combination, *"nat,"* and ending with a more complex, three-syllable pseudo-word, *"l'depnonle.l."* Let me take a moment to pronounce each of these pseudo-words for you as you follow along with Figure 9, beginning with sample item A: /nat/, ib/, /tif/, /hap/, /nan/, /mel/, /jaks/, /lek/, /then't/, /chur/, /fep/, /wus/, /sham[b'l]/, /yash/, /mib[gus]/, /splaunch/, /sast/, /rouch/, /noink/, /kwog/, /lin[di][fi]/, /wum/, /fi/, /hud[ned]/, /me[fret][sun]/, /siTh/, /koj/, and /dep[non][lel].

Using the Word Attack subtest, I gathered information from 135 fourth, fifth, and sixth graders enrolled in public schools located in the Bay Area of Northern California. As I mentioned earlier, all of the youngsters were separated into three different reader categories: *"good"* readers; *"dyslexic"* poor readers; and *"garden-variety"* poor readers. Each of these reader groups was identified in the following manner: 1) *"good"* readers—performance on a standardized scholastic reading test at or above the 50th percentile; 2) *"dyslexic"* poor readers—performance on a standardized scholastic reading test at or below the 25th percentile, and identification as having a learning disability by a school-based student study team; and 3) *"garden-variety"* poor readers—performance on a standardized scholastic reading test at or below the 25th percentile, and no identification as having a learning disability. Since a *"discrepancy method"* between cognitive functioning and academic performance is the most used criterion for learning disability identification in the United States, it was assumed that *"dyslexic"* poor readers met this criterion while *"garden-variety"* poor readers did not.

As expected, according to the equation offered by Gough and Tunmer, good readers were able to decode the majority of pseudo-word items. The *"dyslexic"* poor readers and the *"garden-variety"* poor readers,

Basel . . . . Item 1
Ceiling . . . . 5 consecutive failed

A	________	nat
b	________	ib
1	________	tiff
2	________	hap
3	________	nan
4	________	mell
5	________	jox
6	________	leck
7	________	then't
8	________	chur
9	________	feap
10	________	wuss*
11	________	shomble
12	________	yosh
13	________	mibgus
14	________	splaunch
15	________	saist
16	________	wroutch
17	________	knoink
18	________	quog
19	________	lindify
20	________	whumb
21	________	phigh
22	________	hudned
23	________	mafreatsun
24	________	cythe
25	________	coge
26	________	depnonlel

ADMINISTRATION DIRECTIONS

"I want to read some words that are not real words. Tell me how they sound."

Point to 'nat.'
"How does this word sound?"

Point to 'ib' and say:
"How does this word sound?"

Proceed to item 1 for all subjects.
"Tell me how each of these words sound. Don't go too fast."

Point to each word if necessary. If the subject does not respond in a few seconds, encourage a response. If the subject still does not respond, continue the test y pointing to the next word. Do not pronounce any words for the subject.

Figure 9. Above left, items from the Word Attack subtest of the Woodcock-Johnson Psycho-Educational Battery, Part Two; right, administration directions for the Word Attack subtest.

in contrast, experienced difficulty with the subtest, mispronouncing a significant portion of the *"nonsense"* words. At least at a quantitative level, no differences were noted in the performance of both poor reader groups on the Word Attack subtest. A qualitative analysis of the errors made, however, revealed important information regarding potential differences in the way in which *"dyslexic"* and *"garden-variety"* poor readers attempted to decode the pseudo-words.

In general, when faced with a new *"nonsense"* word, the *"garden-variety"* poor reader would attempt to apply the phonological knowledge they possessed—albeit minimal compared to their *"good"* reader counterparts—in an attempt to pronounce the unfamiliar words. For example: Item #23, "mafreatsun," which is correctly pronounced /[me][fret][sun]/, was pronounced as /[ma][ferts]/ by *"garden-variety"* poor readers. Essentially, *"garden-variety"* poor readers would attempt to employ their decoding skill to read this *"nonsense"* word; but because of their inadequate knowledge of phonics, *"garden-variety"* poor readers fail to look at the entire word, leaving off endings—in this case the /[sun]/ in "mafretsun." Also notice that the internal configuration of the pseudoword, "freat" has also been transposed by the *"garden-variety"* poor reader, pronounced instead as "fert."

In comparison with the performance of the *"garden-variety"* poor readers, which was predictable considering their inadequate knowledge of phonics, *"dyslexic"* readers, surprisingly, pronounced several of the pseudo-words in an entirely unexpected fashion. Since all of the items on the Word Attack subtest are arranged in a hierarchical order, from easiest to hardest, one would expect that both the *"garden-variety"* and the *dyslexic"* groups would reach a *"ceiling of phonics competency"* and begin to mispronounce items consistently. This prediction held true for the *"garden-variety"* group but not for the *"dyslexic"* group. Instead of reaching a *"ceiling,"* youngsters who had been identified as having dyslexia made several unexpected correct responses beyond what would be considered their *"true level"* of performance based on their level of phonological awareness.

Faced with the same *"nonsense"* word "mafreatsun" that was mispronounced as "maferts" by the *"garden-variety"* poor readers, a dyslexic reader used an entirely different strategy to decode the word: Unable to properly decode the word using phonics exclusively, this youngster employed a sight-word strategy, looking for familiar real words embedded within the pseudo-word. This dyslexic youngster read "mafreatsun" in the following manner: He explained that the first part of the word looked like "Ma," the second part of the word looked like "treat" but with an "f" instead of a "t," and the end of the word was like the "sun" in the sky. With each of these parts of the *"nonsense"* word properly classified and recognized, this child was able to put all of the segments together to properly pronounce the pseudo-word. This finding—that *"dyslexic"* poor readers would employ a sight-word strategy in an attempt to recognize real words embedded within a pseudo-word was a consistent one repeated with several dyslexic readers. No *"garden-variety"* readers used this approach to correctly pronounce unfamiliar *"nonsense"* words.

The finding that *"dyslexic"* poor readers might use alternate strategies to decode unfamiliar words might have important ramifications for appropriate educational interventions. If teachers could harness the dyslexic youngster's knowledge of real words or linguistic comprehension, children who have previously experienced frustration could now read words correctly. With dyslexic readers, I also found that several students also got some easier items wrong, whereas the low-achieving poor readers got them right. To be more specific, I found that, when exposed to Item #8 "chur," *"garden-variety"* poor readers would correctly pronounce it, while dyslexic readers would see the word, become frustrated that they couldn't sound it out, and attempt to recall a more familiar real word such as "church." Another example of this phenomenon occurred with Item #26 "depnonlel," the hardest of the items. Attempting to use minimal phonics skills, *"garden-variety"* poor readers pronounced this item as "depnony." Again, one notices that the ending of the *"nonsense"* word has been left off characteristic of the errors this group of youngsters made. They started to sound it out and never finished.

"Dyslexic" readers, on the other hand, unfamiliar with the pseudoword and unable to accurately apply phonics skills, incorporated the "d," "l," and "n" to pronounce the item as "dolphin." Again, *"dyslexic"* readers were relying on their knowledge of real words to compensate for their lack of phonics.

Even though it was clearly explained to all youngsters tested that the Word Attack subtest was a list of new words, not real words, and this disclaimer was repeated during a second administration of the subtest, *"dyslexic"* readers continued to pronounce Item #26 as "dolphin."

Developmental Lags and Phonological-Core Deficits

One researcher has offered a possible explanation as to why this phenomenon occurs; namely, that *"dyslexic"* readers are able to tap into a richer knowledge of linguistic comprehension than their *"garden-variety"* counterparts. In his fascinating book *The Modularity of Mind* (1983), Fodor, explains how the field of *"faculty psychology"* might explain the behavioral differences witnessed in my research. Fodor describes *"horizontal faculties"* as including broad cognitive functions like memory, linguistic comprehension, attention, etc.; *"vertical faculties,"* by comparison, are described as modules of encapsulated information such as phonological awareness, the ability to decode words. Stanovich (1990) has applied Fodor's conceptual framework of *"horizontal faculties"* and *"vertical modules"* to his own theory: Stanovich has postulated that *"garden-variety"* poor readers have a *"developmental lag,"* exhibiting decoding skills and linguistic comprehension comparable to younger *"good"* readers; *"dyslexic"* readers, on the other hand, are believed to have a *"phonological-core deficit,"* exhibiting decoding skills similar to younger readers, but also showing higher levels of linguistic comprehension when compared to their younger counterparts. According to Stanovich's interpretation of Fodor's theory, dyslexic children have intact *"horizontal faculties,"* thereby allowing them to accurately interpret the world around them, understand language and discourse, etc. However, Stanovich also explains that one of the *"vertical modules,"* the one that encapsulates knowledge of phonics, is either missing or significantly diminished in the *"dyslexic"* reader. Stanovich explains that *"garden-variety"* poor readers are not missing the vertical module that encapsulates phonological knowledge but, instead, cognitive functioning including this *"vertical module"* are diminished across the board.

Can Dyslexic Youngsters Become Adult *"Garden-Variety"* Poor Readers?

If it is true that dyslexic youngsters experience a specific weakness in their ability to decode words, then it is also possible that such a localized deficit, over time, might result in broader cognitive deficits. Walberg (1984) has used the concept of *"Matthew effects"* to explain how such a phenomenon might occur. This concept is derived from the Gospel according to Matthew: *"For into every one that hath shall be given, and he shall have abundance; but from him that hath not shall be taken away even that which he hath"* (XXV:29). In other words, Walberg has interpreted this passage as signifying that *"the rich get richer; the poor get poorer"* when it comes to academic gains achieved in the schools. Applying this concept to the area of dyslexia, Stanovich (1988) explains that, if left unchecked, a severe reading disorder ala that experienced by *"Dysphonetic"* dyslexics might result in decreased cognitive functioning. Extending this concept to the different types of readers I have been describing tonight, it is not impossible, as Stanovich (1990) has pointed out regarding Figure 8, that a young *"dyslexic"* reader might become a *"garden-variety"* poor reader over time.

INTERVENTIONS AND TEACHING TECHNIQUES

One of the best interventions for children with dyslexia is working with them on their decoding skills, helping them pick up information to help them circumnavigate, or get past, some of the problems they have sounding out words. It's not as easy as that, because some dyslexic children are going to learn best using a sight-word system instead of a phonics system to teach word identification. By ignoring the fact that dyslexic children access words in different manners (refer back to Figure 8), one might believe that reading problems could be cured just by teaching phonics. About seven or eight years ago, this was the educational method of choice, not recognizing that some children will experience difficulty learning to read using this approach solely.

By not recognizing reading differences and employing alternate educational strategies to teach reading, a disturbing scenario unfolds: A child has been identified as having a decoding problem, in spite of the fact that he/she also exhibits high levels of cognitive functioning, receives instruction in phonics, regardless of the fact that he/she learns faster using a sight-word approach. As a result of using a less effective teaching strategy, this student begins to experience a delay in his/her reading compared with a comparable school-age peer group. As this student gets older, the gap between he/she and his/her peers gets wider.

It should also be pointed out that since students receive a considerable amount of information via reading in the classroom, if such a gap were to develop, widening incrementally over the years it might be highly likely that an individual's understanding of language, and the world in general would be severely restricted. Essentially, for children with reading problems, knowledge of the world starts dropping off because these youngsters are not able to access information in the same way *"good"* readers do. So, a child's educational prognosis, if he/she has a reading problem and if he/she doesn't receive early intervention, gets worse over time.

What, if any, implications do the work of Stanovich and Fodor have for the front-line teacher? First of all, it is important to point out that the classroom teacher will be sensitive to individual differences in reading among his/her students, and it is this awareness of reading differences that proves to be the single most important factor in teaching *"dyslexic"* and *"garden-variety"* poor readers. A teacher must recognize that reading skills can be mapped on two axes (see Figure 8): phonic reading, and whole word reading. All youngsters will have some combination of these approaches to reading the printed word. Once one recognizes that not all children have excellent phonic and whole word reading capabilities, it is quite easy to design curricula to address the learning style of individual youngsters instead of forcing all poor readers into a phonics approach.

Secondly, the reading equation proposed by Gough and Tunmer indicates that decoding is only one of two important variables, each making a significant contribution to reading comprehension; the other variable is linguistic comprehension. Teachers should be giving each of these three variables equal attention: Children need to be able to decode unfamiliar words; they should have sufficient linguistic comprehension to make sense of the world around them; and, they should be able to apply both decoding and linguistic comprehension to make sense out of the material that they read. Without sufficient attention to these factors, teachers will, in many cases, witness *"Matthew effects"*; Students will learn less and less since their main avenue of receiving information, reading, will have effectively been limited.

For the *"dyslexic"* and the *"garden-variety"* poor reader alike, teachers must offer direct instruction in phonics as well as providing an enriched environment—one that encourages new experiences and is filled with language via discussion, conversation, and presentations. Let's focus on the first part—phonics. To be more specific, teachers can approach decoding instruction from one of two directions: (1) *"meaning emphasis"*; (2) *"code emphasis"* approach. The *"meaning emphasis"* approach involves exposing children to language in books and words so that slowly they begin to acquire a whole-word vocabulary. In the *"code emphasis"* approach, phonics is taught directly, waiting to teach word meanings later. By and large, the *"code emphasis"* approach is taught in one of two manners. The *"Analytic Method"* exposes children to words that are similar except for one phoneme that has been changed (e.g., "pat," "mat," "fat"). Another way of teaching phonics, using the *"code emphasis"* approach is known as the *"Synthetic Method,"* whereby separate phonemes are taught in isolation and are later blended together to signify a whole word (e.g., "s," "a," "t" are taught first; then they are blended to become "sat").

Although the *"code emphasis"* approach has been the educational technique of choice for assisting poor readers, current research (Bryant & Bradley, 1985) indicates that children don't learn how to decode words by breaking them into phonemes (Analytic Method) or by blending phonemes back together (Synthetic Method). Some children might benefit from these techniques, but, by and large, children learn different phonemes by rhyming. Bryant and Bradley (1985) report that, instead of learning sounds phoneme by phoneme, children are sensitive to two parts of a word: the "onset," the beginning phoneme, and the "rime," the rest of the word. For example, in the word "nest," the "onset" would be the phoneme "n," and the "rime" would be the rest of the word "est." In short, Bradley and Bryant propose that children eventually come to understand the importance of phonemes by discriminating between words that begin with different phonemes but end with the same "rime" (e.g., "nest," "best," "test"). Bryant and Bradley continue to explain that it is this understanding of "onset" and "rime" that allows a child s phonological awareness to emerge; youngsters become sensitive to phonemic differentiations and slowly recognize that these phonemes have distinct identities.

As a result of this discovery, Bryant and Bradley (1985) have developed a powerful new educational technique that consists of teaching children how to rhyme words. Bryant and Bradley base this intervention on their findings that children who experience difficulty rhyming early on, have a higher probability of encountering decoding and, hence, reading problems later in life. On the other hand, children who are good rhymers were found to become good readers later on.

What children's book would be an excellent way of teaching rhyming? Dr. Seuss, of course! I have recommended Dr. Seuss books to parents of youngsters, regardless of whether their children are experiencing reading difficulty or not. Dr. Seuss books accomplish two objectives: First, they teach rhyming in a fun and novel manner; second, they employ *"nonsense"* words, allowing children to explore their understanding of phonics without confusing this understanding with common, high frequency real words. If one can teach youngsters to *"play"* with language by using *"nonsense"* words, children can develop a keen sense of phonics.

SUGGESTIONS FOR PARENTAL INVOLVEMENT

Now, I would like to offer some advice to parents of children who are experiencing reading problems. First, it is extremely important for parents to read to their children, especially at an early age. Get them the information that they're being cut off from in school. Read to them, expose them to different experiences vis-a-vis trips to museums, national parks, zoos, etc. If parents have a young child that they suspect might have dyslexia, they should expose their youngster to *"rhyming games."* As I have already pointed out, one excellent source of rhyming can be found in the Dr. Seuss books.

Secondly, actively encourage discourse in the home through conversations with adults and between siblings. Remember, that parents must present information in novel ways, since the printed word is so difficult to decipher. Parents should make sure, whenever the opportunity presents itself, that their child's world is filled with language—whether it be dinner conversation, exposure to new vocabulary, etc. When I make this suggestion, I don't mean showing flash cards but, rather, providing activities that are language oriented.

By reading to children and actively engaging them in conversation, parents can keep their children intellectually and educationally afloat as they go through a potentially frustrating—and consequently, emotional—period of their lives.

These suggestions, so far, pertain to the younger dyslexic child. What happens to the dyslexic adolescent who has already been in school for several years, not receiving the interventions I have mentioned? Do parents focus on phonics exclusively? The answer to this question is "no."

As the dyslexic child becomes an adult, increasing emphasis should be placed into the provision of appropriate accommodations, so the dyslexic adult can access information in spite of his/her poor reading skills. For someone who is graduating from high school, it is important to provide tools such as *"books-on-tape"* that will allow the dyslexic adult access to important real-world information.

During my tenure at the Disabled Student's Program at the University of California at Berkeley, many of the dyslexic students that I worked with complained that they didn't want to read textbook chapters, because they had to read word-for-word an extremely laborious and frustrating task. With *"books-on-tape,"* however, these very same students stayed on top of their assignments, learned the requisite information, and went on to earn an undergraduate degree. That's not to suggest that dyslexic adults shouldn't be taught how to increase their decoding and comprehension skills; this should be a lifelong endeavor, or at least until the dyslexic adult reaches a reading level that he/she is comfortable with.

CONCLUSION

One last point I wish to make is that dyslexia is not a disease like measles. If one has to think of dyslexia as a medical phenomenon and it most surely involves neurological functioning one should think of it as analogous to obesity (Stanovich, 1990). People who are overweight to a certain degree are considered to fall into the category of *"obesity."* Obesity is a medical phenomenon that is researched for potential causes, and interventions are designed to prevent people from becoming obese. Considering for one last time Figure 8, the continuum of good and poor readers, one must acknowledge that statements such as, *"This child is dyslexic"* and, *"This child is not dyslexic,"* fly in the face of all of the research I have presented. Dyslexia, like being overweight, should be thought of as a real, albeit relative phenomenon. All individuals have different reading strengths and weaknesses; some people have excellent whole-word reading skills but poor decoding skills; others have excellent decoding skills but poor whole-word reading skills. Because dyslexia has become an *"either-or"* phenomenon, many children are not receiving the instruction they require since they have not been identified as *"dyslexic."* And many children originally identified as having dyslexia are denied services later in life when they no longer qualify as having a *"severe"* reading problem.

All of the information that I have presented tonight has given us some insight into how children learn to read; this information should be an integral part of their education for the rest of their lives.

QUESTIONS AND ANSWERS

Question: Do you see movement within the continuum depicted in Figure 8?

Dr. Spagna: This is an important question, and now I'm going to focus on the New York Times article I referred to earlier in the evening. The article reports a study out of Yale University that concludes that dyslexia is not a life-long phenomenon. You can see on the continuum of reading skills depicted in Figure 8 that, maybe, if certain skills improve say whole word reading skill, for instance—a child might move out of one quadrant and towards another. The Yale research group has already addressed the question by stipulating that, eventually, some children experience skill development. Unfortunately, instead of stating that poor readers might experience some gain in phonic or whole word reading skill, the New York Times article left the public with the impression that dyslexia was "curable."

Dyslexia is not a medical condition that is "cured." A better way to describe dyslexia is to emphasize that all individuals experience differences in their ability to apply phonic and whole word reading skills when attempting to derive meaning from text. It should be expected that children who receive direct instruction in phonic reading skill should experience improvement and, hence, would move to a slightly different position along the reading skill continuum.

Question: I saw something in *Scientific American* which pointed more to the biological aspects of dyslexia. Is there any truth to that?

Dr. Spagna: I would say that the whole area of brain research is still in its infancy and we definitely still have a way to go in this area. As I explained in the modularity theory, we don't know why dyslexic children are missing this one vertical module while all of their horizontal faculties are intact. The hard thing to decide is whether something happened to the brain early and therefore the module was lost, or did the child lose the module and then the brain didn't develop the same way? We still don't have enough knowledge of neurology and brain function to be able to figure out this "chicken-and-egg" debate. I think we need to know if there are any types of biological precursors to the development of phonological awareness. For instance, Galaberta, at Harvard, and other people, are doing research that I would fully encourage. Without this research, we will never determine the etiological reasons for dyslexia.

Question: How can you test for dyslexia if these "Matthew effects" actually exist and if the child has poor reading skills and later on has a cognitive deficit as a result of this? Should we test the student later in life?

Dr. Spagna: If you think about intelligence tests, they are language-loaded. So, a child who might have language deficits as a result of his/her reading problem, because he/she has inadequate phonological skills, will do poorly on an intelligence test. Is that child going to be diagnosed dyslexic? NO; because they don't meet the cut-off. They might have had a big gap between aptitude and achievement if they were tested when they were young, but as they got older, the discrepancy no longer exists. So, in this instance, I would say that, especially if it's an older learning disabled individual, you would have to really ask, "What is the intelligence test testing?" And, "Is that the best method for determining aptitude?" I'd probably say, "no." You would have to go back into the child's history to determine whether early precursors were present: Were there teacher reports that showed that the child had difficulty reading at an early age? Did the parents notice early signs of reading difficulty? Was language acquisition later than normal?

Then, you can side-step the state eligibility criterion—that the child has to exhibit a significant discrepancy between aptitude and achievement—and consider school history, family history, medical history, etc. as an indicator of reading problems. You can act around the pure discrepancy definition of dylexia by looking at other sources, primarily family and educational background.

Question: You didn't say anything about writing. Are you dissociationg that from what we're talking about tonight?

Dr. Spagna: No. Quite contrary. I think reading and writing are very much intertwined, but I just didn't have the time tonight to talk about it. I think many children learn to read via their experimentation with the written word.

Question: How important is the emotional or motivational attitude of a learner?

Dr. Spagna: There have been quite a few studies, especially in the area of "Attribution" theory which are relevant to learning and motivation. Younger children are not as upset when they encounter difficulties in reading. Reading is a new phenomenon, so they're willing to make mistakes. Later on in life, though, as these children move from elementary school to junior high school, they start adding emotional reactions to the fact that they can't read. There's an excellent book by Seligman which is called *Helplessness*. He talks about the classroom and mentions that, if students are not being stimulated by their educational environment, sooner or later they'll "turn off" to all education. Now, I would have to agree with Seligman's assessment. I have met adults with learning problems, who didn't have their problems addressed, who won't pick up a newspaper or read because it's so painful for them emotionally.

REFERENCES

Baron, J., &Strawson, C. (1976). Use of orthographic and word-specific knowledge in reading words aloud. *Journal of Experimental Psychology: Human Perception and Performance, 2*, 386–393.

Berger, Yule, & Rutter (1975). Cited in R. M. Knights & D. J. Bakker (Eds.), *The neuropsychology of learning disorders*. Baltimore: University Park Press.

Boder, E. (1973). Developmental dyslexia: A diagnostic approach based on three atypical reading-spelling patterns. *Developmental Medicine and Child Neurology, 15, 663–687.*

Bryant, P., & Bradley, L. (1985). *Children's reading problems*. New York: Basil Blackwell.

Coltheart, M., Patterson, K., & Marshall, J. C. (1980). *Deep dyslexia (2nd ed.)*, New York: Routledge & Kegan Paul.

Ellis, A. W. (1985). The cognitive neuropsychology of developmental (and acquired) dyslexia: A critical survey. *Cognitive Neuropsychology, 2*, (2), 169–205.

Fernald, G. (1943). *Remedial techniques in basic school objectives.* New York: McGraw-Hill.

Fodor, J. A. (1983). *The modularity of mind.* Cambridge, MA: MIT press.

Gough, P. B., & Tunmer, W. E. (1986). Decoding, reading, and reading disability. *Remedial and Special Education, 7,* (1), 6–10.

Hinshelwood, J. (1917). *Congenital word-blindness.* London: H. K. Lewis.

Kirk, S. A. (1962). *Educating exceptional children.* Boston: Houghton Mifflin.

Orton, S. T. (1928). Specific reading disability—strephosymbolia. *Journal of the American Medical Association, 90,* 1095–1099.

Public Law 91-230 (1969). *Children with Specific Learning Disabilities Act.* Washington, DC: U.S. Department of Health, Education, and Welfare.

Public Law 94–142 (1975). *Education for All Handicapped Children Act.* Washington, DC: U.S. Department of Health, Education, and Welfare.

Rutter, M., & Yule, W. (1975). The concept of specific reading retardation. *Journal of child Psychology & Psychiatry, 16,* 181–197.

Sabatino, D. A. (1981). Overview for the practitioner in learninf disabilities. In D. A. Sabatino, T. L. Miller, & C. Schmidt, *Learning disabilities: Systemizing teaching and service delivery,* 1–24. Rockville, MD: Aspen.

Shaywitz, S. E., Escobar, M. D., Shaywitz, B. A., Fletcher, J. M. & Makuch, R. (1992). Evidence that dyslexia may represent the lower tail of a normal distribution of reading ability. *The New England Journal of Medicine, 326,* (3), 145–150.

Shepard, L. A., Smith, M. L., & Vojir, C. P. (1983). Characteristics of pupils identified as learning disabled. *American Educational Research Journal, 20,* (3) 309–331.

Stanovich, K. E. (1990). Explaining the differences between the dyslexic and garden-variety poor reader: The phonological-core variable-difference model. In J. K. Torgesen, (Ed.), *Cognitive and behavioral characteristics of children with learning disabilities,* 7–40. Austin, TX: Pro-Ed, Inc.

Stanovich, K. E. (1988). The right and wrong places to look for the cognitive locus of reading disability. *Annals of dyslexia, 38,* 154–177.

Warner, M. M., Schumaker, J. B., Alley, G. R., & Deshler, D. D. (1980). Learning disabled adolescents in the public schools: Are they different from other low achievers? *Exceptional Education Quarterly, 1,* 27–36.

Wiederholz, J. L. (1974). Historical perspectives on the education of the learning disabled. In L. Mann & D. Sabatino (Eds.), *The second review of special education,* 103–152. Philadelphia: JSE Press.

Woodcock, R. W., & Mather, N. (1989). *Woodcock-Johnson Psycho-Educational Battery–Revised (examiner's manual).* Allen, TX: DLM Teaching Resources.

Ysseldyke, J. E., Algozzine, B., Shinn, M. R., & McGuire, M. (1982). Similarities and differences between low achievers and students classified learning disabled. *Journal of Special Education, 16,* (1), 73–85.

LECTURE 6

A. Martin Goodman, Ph.D.
Clinical Psychologist

FAMILY DYNAMICS AND LEARNING DISABILITIES

*"You never pitied me, yet I felt you understood my struggles.
I can have my doubts and at the same time still feel strong."*

Introduction by Richard L. Goldman

As we continue our speaker series, we are fortunate to have as our speaker tonight an individual who is a clinical psychologist in private practice and a trained psychoanalyst. For twenty years, Dr. Goodman has specialized in individual therapy with children, adolescents and adults as well as couples and family treatment. He is currently the Consulting Psychologist for Landmark West School, serves as a clinical supervisor and was previously Director of the Adolescent Drug Treatment Program sponsored by the National Institute of Health.

In addition, through the years he has conducted many presentations, seminars and university lectures.

Dr. Goodman has extensive experience working with individuals who have learning disabilities and attention deficit disorder. Of particular interest is the subject of the inner world experience of these people and their families. Tonight's presentation will address this concept as the foundation for the development of self esteem and self advocacy.

To address "Family Dynamics and Learning Disabilities," Dr. Martin Goodman.

Dr. Martin Goodman

Family dynamics and learning disabilities is a demanding topic for all of us. Being a parent under the best of circumstances is a difficult task. Certainly when there is a young person in your family with a learning disability, parenting is even more difficult. Many times we feel helpless and frustrated and, in these circumstances, often the best of who we are does not come through. The purpose of this evening's presentation is to explore and understand not only the experience of the LD child, but that of his family as well.

Those of you who are here for professional interests, and who do not have LD children, will also find this information useful. Insight into the parenting process is valuable for all of us. We all need to work at parenting. We put so much energy into other aspects of our lives (at the workplace, for example), but we often fail to realize that parenting is a process requiring particular skills needing to be cultivated. We have often heard the saying, "It is easy to become a parent, but being one is very difficult." In the course of this evening's discussion, as we explore thestruggle with the parenting process, I will add the following: *"All parents make mistakes—mistakes are an integral part of the learning process. Probably he who never makes a mistake never makes a discovery."*

Our goal is to learn how to be an advocate for the LD child and to help him or her become an advocate for themselves. All of these children have numerous strengths and abilities which we draw upon to aid in this process. However, tonight I will be focusing on some of the internal struggles these children have. You may at times say to yourselves, *"That's not my child"* or, *"My child is not that bad off."* But to some degree your child will share some of the characteristics I will be describing this evening.

When we think about your child as an advocate for him or herself, the most difficult part is overcoming their feelings of being alone with their struggles. We want to turn the feelings of defeat into experiences in which young people can gain an internal voice that says, *"Hang in there, it's O.K., I'm with you."* If they do not have an internal voice that comes through family experiences, then they will feel isolated. I want to convey to you the LD child's internal emotional tone. In our search for this inner feeling, it is essential that we understand our young people and their vulnerabilities.

I want to convey how we may hear their struggle, so they will not feel alone in this life. It is bad enough that a youngster has a learning disability, and many times the learning disability will persist. It is crucial that we help them develop a depth of self-understanding out of which self-esteem will emerge so they can effectively cope with their difficulties.

Learning disabled children embody a broad range of characteristics. Some have almost imperceptible learning disabilities and are intact in many ways. Others have a pervasive learning disability. We find that many LD children have varying levels of performance., so that while they may have exceptional capacity in some areas, they may be significantly deficient in others. What we want to do is build self-esteem that enables them to say, *"Well, O.K., I will have obstacles and embarrassments because of my learning disabilities, and some of these I cannot overcome. But if there is one thing that my school and my family has taught me, it is how to hang in there and still believe in myself."* We want to help them gain the internal as well as external skills to persevere in the face of adversity. After all, if all of us could do this in life, we wouldn't be doing so badly either.

Learning disabilities are seen in both active and passive forms. Sometimes we see attentional deficits which are active and the youngster is acting out all over the place. We also see the passive child. This second type of child is withdrawn, very shy, and presents a very vulnerable picture. The second youngster does not want to take on anything new for fear that he will experience another defeat. His or her daily life is often in the form of crisis.

In my clinical psychotherapy practice, I frequently ask, *"What's new this week?"* However, in most people's lives, things do not change much from week to week. But with LD kids this does not hold true, for there is always something new. Every single day is filled with difficulties. They cannot sit still, they daydream, and they are often seen as troublemakers because they lack the capacity to constructively verbalize their anger and frustrations. As a result of this, they are often in a state of crisis.

We want to assist these young people with the communication and articulation of their feelings. It is very difficult for them to transform internal experiences, feelings and thoughts into language. What we will try to do this evening is to focus on how to help them so that they can bring what is inside to the outside. This is a very therapeutic thing to do.

Learning disabled children cannot easily read social cues and do not know how to fit well into social groups. Psychological maturation is learning about the shades of grey in social situations. For example, an adolescent who is not yet mature psychologically, like an adult, has the tendency to see the world as black and white. Little children will say simplistically, *"Is he good or is he bad?"* However, as we progress through life, gaining wisdom, most of us are able to determine shades of grey and subtleties within our social context. However, the learning disabled child does not grasp that part of social processing and does not know about shades of grey. They do not know to enter a social situation in which they must read the many subtle cues and then work their way into the setting. Thus, many times we are working with a lonely child, one who stops trying socially because he anticipates defeat. Everyday activities for LD children can thus become crisis experiences. If most of us are on the periphery of a group occasionally, this is O.K. But if this is your daily experience, if others say things that you don't understand and continually humiliate you, daily activities become continual crises.

This is especially true for some types of LD adolescents. If you could go back to any age of life, how many of you would like to go back to the seventh grade? I do not think many of you would choose to do so. It is a time during which there is much turmoil.

We will talk later about the LD adolescent, because this is an area of special needs and it is a particular interest of mine. However, for the moment, let us continue with the daily mini-crisis of the LD child.

For example, the child with both an attention deficit and a learning disorder who leaves home without his lunch experiences mini-crisis. Even though you remind him a dozen times, he still forgets. Even though he forgets, it does not mean that he does not care and, as a result, he often feels ashamed. Another example of a situation like this would be when confronting a difficult exam. The child may not only feel that he did not succeed, but also may feel deep shame because of repeated failure. It is one thing to do poorly in an examination, but imagine this failure after you really studied. Other people will often think you are lazy, even though you know you tried very hard. Think of your own experiences. Since some of you may be learning disabled also, you can become an ally to your son or daughter. The fact that you were learning disabled may prove to be an asset. If this is true, then you might become better able to get in touch with your son or daughter. Search yourself and try to remember what it was like to be that age. Now, picture yourself as being an LD child studying for two hours before an exam and then, just when you think you might have conquered the exam, you get it back with a grade of 'D'!

The child might say, *"They always tell me I'm not stupid,"* but deep down, the child feels stupid. The last thing we want to have happen is for that child to carry a sense of defeat and shame in an isolated fashion. If the feeling remains isolated, and cannot be understood by others who are able to reach out and to feel that disappointment also, the child will simply wall off the feelings, and the feelings will become part of an accumulation of shaming experiences. This will eventually turn into what is referred to as *"chronic shame."* This is the kind of shame that occurs when someone comes in contact with certain experiences and shame becomes overwhelming. The LD child thus starts developing a phobic avoidance toward academics.

A little bit of shame is common for all of us. We all cannot do everything well; all of us will be better at one thing or another. However, chronic shame to the self is the type of shame that builds up and says, *"I am a poor excuse for a person."* The shame is not centered around an act; it is centered around the identity of the person. This kind of shame is part of the everyday life of the LD child.

If you see that type of shame displayed in your child, do not deny or dismiss it. Rather, be understanding and sensitive. Even though LD children may look very happy and may be doing O.K. in other ways, they feel shame because of all the secrets they hold inside. We try, of course, to help remediate their difficulties and sometimes we can be very successful, particularly in schools like Landmark. One of the remarkable opportunities the child has here is not only to have things targeted for remediation, but also to obtain sensitivity from the staff. If something seems overwhelming, the staff has a way of breaking the problem down into manageable parts so the child does not feel desperately inadequate. It is a way of reaching out to them and it serves both an educational and an emotional function.

I am focusing on a whole range of emotions common to LD children, including the feelings of shame, inadequacy, low self-esteem, hopelessness and helplessness, all of which LD children experience. They are vulnerable to insults from peers and from siblings. In the family, siblings can be quite mean. When they get really angry, they can hit below the belt. The LD child becomes shamed and stores unresolved problems internally. Their reaction may be tremendous rage, which is a byproduct of shame reactions, or they may just give up, defeated, saying to themselves, *"The hell if I will expose myself to further defeat or embarrassment."*

In the past, your child may have had teachers who did not understand their learning disabilities. These teachers might have said, *"Come on, you can do it,"* and maybe even meant well but, in essence, they shamed them because they did not really understand. It was not that the child was not trying, but that they really needed special help to solve the problems.

Parents may also be involved in producing these self-esteem difficulties. It is not uncommon to have one parent more convinced about the presence of a learning disability, while the other parent thinks that, *"The child is just lazy."* It is a very difficult situation in which the LD youngster may find himself. We all want our children to succeed in this world. We are afraid for our children and, as parents of LD children, you must decide to what degree do you push them, and to what degree do you accept their deficit? This type of question is better answered when we understand both the educational as well as the emotional realities. There are ways to encourage young people when we have genuine insight into their special learning problems, as well as to the feelings surrounding their difficulties.

ACTIVITIES OUT OF THE HOME

Finding appropriate activities out of the home is difficult. LD youngsters may have few friends. They also have many problems with poor judgement. Once again, they have difficulty with the various social cues, the contextual meanings that involve shades of grey, and the ability to think things through like their non-LD sibling. And so, parents may often get furious with them. Parents will state with anger and exasperation, *"I thought you knew better. How could you have done that!"* Many negative acts by the LD child are simply ways of communicating their frustration.

Many of these children who have difficulties in articulating their feelings and thoughts often get in trouble because they do not have any real way of saying what they need to say. They have difficulties knowing why they were angry or why things were unfair or unjust. Instead of articulating, they retaliate and do things which are equally unjust. They will seem to be saying by their actions, *"You see what it feels like? I'll show you—try this one on for size!"* Consequently, they get into more trouble. When they are sent to the office for behaviors like this, they sit there with self-righteous indignation. What they are really feeling inside is, *"Did you see what happened to me, how unfair things are?"*

In addition, learning disabilities directly relate to study habits. When they look at their books, they think about how inadequate they are. One of our tasks as parents and educators is to help young people deal with these feelings of inadequacy and learn how to resolve and tolerate these feelings so that they will not sit at their desks with their books shut. One of my tasks as a consultant at Landmark is to work with the staff and help them form links and connections to the young people. In the tutorial, or prep program, we try to help the staff prepare students for life itself.

Every child wants to feel special. But how do LD children feel good about themselves when others, as well as their siblings, perform more adequately than they do in many ways? One of the common things we say about LD children is, *"Let's help them by building upon their strengths."* But with some LD children, it is sometimes difficult to find their strengths compared to those of their siblings. We really have to look for strengths, and this requires special skills and sensitivities.

SIBLING PROBLEMS

Let's talk about the interactions between LD children and their siblings. The child often has a lot of resentment directed toward their siblings. Siblings are perceived by them as perfect. It is a natural experience for an LD child who is experiencing defeats in life. The youngster might feel deeply conflicted about feeling resentment towards competent siblings, because they really like their brother or sister. The able sibling may be working very hard to succeed, but have difficulty understanding that. Because there are so many problems, he often gets the lion's share of parental attention.

Thus, there are many mixed feelings that the siblings have towards their LD brother or sister who may be getting more parental time and attention. Many times, besides the simple resentment that they feel, there is a whole host of other feelings that are present. Normal siblings often feel guilt toward their LD sibling. Because they are doing so much better, they can sense and see the pain on the face of their LD brother or sister. The academically adequate child may just waltz in with A's, or be successful in sports, and they feel a little guilty about this because they do not want their brother or sister to feel badly in comparison. However, their guilt is not as strong as their normal desire for positive attention and a need to be special. No child can resist this. Moreover, they are entitled to have the family resonate with their success. They are often angry that they have a brother or sister whom they feel guilty about. Many times, when this anger goes unexpressed, it becomes bottled up and comes out as resentment.

Academically average siblings may also feel guilty because they are often embarrassed by their LD siblings. If they have a brother or sister who, by their standards, is not in the mainstream, they are embarrassed. So these normal siblings may experience turmoil. They also may love someone they do not like. Additionally, they feel pressure to succeed because of the presence of an an LD sibling, and they do not want to let the parents down by being another problem child. Many of them are hurting or struggling or have their own problems, but they see their parents struggling with their troubled sibling and, thus, they may keep all their problems to themselves. In addition, they may also see an older LD sibling having serious problems in the upper grades and think that, "The higher grades must be terrible. Is that what it is like; do you fall apart when you get older?"

There are all sorts of natural feelings of rivalry in every family and the whole phenomena complicates this rivalry. Sometimes rivalry and resulting competition can be healthy in circumstances where there is enough love to go around. The children will push each other and that results in healthy striving. When a LD child is involved, however, the child is vulnerable to feelings of inadequacy and defeat.

Lastly, the competent sibling of an LD child often has to live with the burden of being the object of tremendous envy. They live with the inner sense that another person somehow wants them to fail. (Actually, we all have those feelings, in the internal corners of our inner life.) Thus, when that sibling has a defeat, there is a part of that LD child who is rejoicing. The healthy brother and sister know it. They live with this feeling and it creates pain for them. As parents, it is crucial that we not judge these normal reactions in our children, but respond with a level of sensitivity and assist with understanding so that they may integrate these feelings.

SOLUTIONS

Let's continue to talk about some solutions. I will first go over some ideas as to how to communicate with these children. I will try to deal with the parents' experiences. After this, I will answer some questions and try to develop a dialogue. In a dialogue setting, I can get my clinical orientation out, as most of the time I work in a clinical setting. The only real control we have as parents is over ourselves. You cannot really control your children. You can try to control your child, but it does not work. What I would like you to do is to learn how to control yourself and to be in touch with your emotions and feelings. As parents, your job is to teach your children to become more effective people in this life. Thus, the most crucial thing you have to do is to provide successful models through good communication. This means effective communications about abstract ideas, as well as about everyday life.

One of the big obstacles parents have is to come to grips with the fact that they have significant feelings of loss and disappointment with regard to their LD child. When the child is in the cradle, all parents first think, *"Well, he can be what he wants to be."* When parents see that their child is learning disabled, parents may feel that their dreams for their child have been shattered. You need to be honest with yourselves. Freud said that, *"Children read our unconscious, not our conscious."* In other words, they see our moods, they read our body language, they watch us and study us and see our emotions from their birth. They know at some point how you feel about them and what you think. I suggest to you that unless you search yourselves for your own true feelings, they will come across anyway. So you must get control of yourselves so that you can get a sense of what is coming across to your children.

With any sense of loss, there usually occurs a three step process. These steps involve denial, anger and acceptance. Often, when parents first hear that their child is learning disabled, they first think that the school is wrong, or they just do not know "my child," or they do not know how to work with "my child," etc. This is upsetting and this constitutes a state of denial.

Then anger comes, as this is part of denial. Parents think, *"Let's find a new school!"* The parents still have not accepted what their child needs.

There are a number of parents who have their children here at Landmark that need to look at their own denial process. They have accepted the basics of the children's learning disabilities. However, they still need to acknowledge their own unresolved feelings about this fact. One of Landmark's goals is to give children the sense that they are in as normal a school setting as possible. The staff here wants to give children the feeling that they are part of a large social system, and to accept real life. Yet, on the other hand, these children cannot do all things.

There was a final examination period recently, in which some of the children were *"freaking out"* over their examinations. For some, it was too much; they could not cope. There were other children who did quite well. They needed, however, to learn to cope with this type of situation, as life presents many situations which involve final exams. Life is full of pressure situations, so it is appropriate for Landmark to carefully approach the examination period as an opportunity to learn how to more effectively cope with pressure.

We have parents here at Landmark who sometimes state that, *"My son or daughter has to get a good grade in this algebra examination because, if he doesn't, he will not get into college."* While it is true that parents must face the fact that some LD children will not go to college, there are many learning disabled children who will go to college. My brother-in-law is learning disabled and he majored in his strength, music. However, it took him six years to graduate and it was tough for him. For those who will not make it into college, we have to begin to deal with this possibility and with our own disappointments as parents. We should not project our negative feelings onto our children, and thus make our youngsters feel ashamed of themselves.

For example, if you wanted your child to become an attorney, it is better not to ignore your feelings and just pass them off. It is better to come to grips with your feelings and tell yourself that you are genuinely disappointed with the fact that he will not become an attorney. If you can face these feelings in yourself, you are not as likely to project them outwardly towards your child. Whenever you have a lack of understanding, then you might take the blame. When you have not come to terms with the pain, you will suffer blame. Parents devise a variety of ways of blaming themselves. By blaming themselves, they feel guilty. The parent may think, *"I should have spent more time with him of her,"* or, *"It must have come from my side of the family."* Thus, it is not uncommon for this to create a lot of family discord. If you have, to cite another example, a hyperactive child, you experience difficulties from the earliest stages of his or her life. They may be up every half hour during the first six months of their lives, and this can drive parents nuts. Parents may then begin to disagree energetically on the parenting style to use, to be too tough or too lenient. There is usually pain and something needs to be done about it. When you do not understand it, you will get into a cycle of blame. Sometimes parents will compensate by making their child's problem their reason to live. Everything centers around the child with the problem. But that is not really what we want to do with these children. We want to give them balanced social experiences.

So, now we have all these problems. What are we going to do about them? The answer is simple. Parents should try to become a conduit for their children, and help them express their inner feelings outwardly. They can express what is inside with our assistance as parents. We should not try to shut them down. We should teach them how to deal with a whole host of emotions. We want to help them through what is called *"reflective listening."* Reflective listening works like this: You think of yourself as mirrors, so that whatever your child says to you, you then repeat it back to them as a mirror would. Two books are very good in finding out more about reflective listening. They are: *Parent Effectiveness Training* by Thomas Gordon and *Step Systematic Training for Effective Parenting* by Don Dinkmeyer and Gary B. McKay.

Reflective listening thus involves stating the child's feelings and meanings so that he or she feels that you really hear and understand them. Every child, particularly teenagers, desperately needs to be understood. In doing this, we will help them achieve better self-control. We want to say to ourselves, *"What is my child feeling, and what brought on this feeling?"* For example, if your child comes home from school and has had a fight with the teacher, typically the parent wants to get into the problem immediately by asking, *"What did you do, what did the teacher do?"*, thus entering the process of assigning blame. What you really need to do, however, is to help your young person sort the problem out. You don't want to appear to take sides. You just want to help them reflect back upon their unpleasant experience.

For example, if your child says, *"That teacher was so mean to me, or, I hate that teacher,"* the parent should then, rather than assigning blame, say, *"Boy, you really sound angry,"* which invites the child to tell you a little bit more. The child might then say, *"Yeah, that teacher was really unfair to me!"* Sometimes, at this point, the parent may be tempted to say something that will shut the child down, such as, *"Sometimes you really do misbehave."* Instead, you might say, *"You feel the teacher was really unfair. Tell me about it."* Thus, you are not taking sides and you are not deciding that either the child or teacher was wrong. You are thus inviting the child to articulate more. The child may then say, *"Well, that teacher singled me out and sent me to the office."* It would be tempting at this point for the parent to say, *"You have been sent to the office so many times, you must be doing something wrong."* But, keeping the mirror concept in mind, you can say, *"You were sent to the office, that sounds really upsetting. What happened?"* Now this child still has not been shut off, and is not arguing, and will likely tell you what happened. He might then say, *"Yeah, everyone else in the class was talking and everyone should have been sent to the office, but she just sent me to the office."* The reason I am using this example is that, within my practice recently, I had a child who was constantly testing the limits. For example, when he was asked by his teacher to be quiet, he would immediately whisper to a neighbor and the teacher would say, *"Go to the office."* It would be very easy to blame him and to make him defensive. Keeping the mirror concept in mind, I said to him, *"You felt you were being singled out. It must have been terrible."* He responded with, *"Yes, it was terrible; I hated it. That teacher had it in for me."* This type of communication can give you an inroad into the child. You may continue with, *"You know, sometimes you are really angry with your teacher, aren't you? But you have also told me that he is not so bad. What do you think happened in this situation?"* If you keep him in this reflective mode, you then begin to have a link to the child. The child may respond by saying, *"I have been singled out."* You might then respond with, *"Yeah, I hate to be singled out, and sometimes I was, and it was terrible."* Once again, you have continued to build up a link; you have held up a mirror. At this time, it could be time to work on a solution such as, *"So, now what do you think we might do to solve this problem?"* If the child says, *"It is hopeless; it won't work,"* you might respond by saying, *"I see you feel really frustrated and you are giving up and you do not want to really try to solve the problem."* Once again, he will feel that you are linking with his feelings. You can continue in this mode with, *"Let's think of some good ideas together."* At this point, neither the teacher nor the child is being blamed. Thus, you have built up a link of empathy so that he does not feel that you are putting him down or blaming him. You have not taken a position. You are still in a problem-solving mode. Part of the problem solving mode could involve helping the child identify social cues that communicate when the teacher is really serious. At those times it is essential that no more talking occurs.

Let's take another example of a situation cited in the book *Between Parent and Teenager* by Dr. Haim G. Ginott. An LD child returns home from a job interview. During the interview things did not go very well. The unattuned parent might say, *"What did you expect, that you would get your first job on your first interview? Life is not like that. You might need to go to five or ten interviews before you are hired."* However, if the parent works as a mirror, they could say, *"Gee, I know you were looking forward to this job and I know you are really disappointed."* Thus, you and your child are on the same team. In contrast, the well-meaning but unattuned parent would say, *"Rome was not built in a day, chin up and go get 'em, Tiger!"*

Although this kind of encouragement can be appropriate, if you sense defeat in the child, you might say, *"You look pretty down about this. I get the feeling that you do not want to look for other jobs."* The child might say, *"Yes, that was a terrible experience."* The parent can continue mirroring by saying something like, *"You are really humiliated and you feel like you will never get another job,"* and the child might respond with, *"Yeah, I feel like I will never get another job."* At this point, the parent could say, *"I know you really want to get a job, let's think together how you can approach this problem again."* The parent can forge a link to their internal experience. It does not help to minimize difficult situations by saying, for example, *"I do not see why you are depressed. It was only one job!"* If you do this, they will not only feel discouraged, but they will feel inadequate because they are discouraged.

Admonishing children for not having enough self-esteem will not improve self-confidence. The forming of links is an old therapeutic technique that we use to enhance communication with the child in non-judgmental ways. It can be quite productive. We also do not want to pity the child by saying, *"Oh, you poor thing. What terrible luck."* He is not a defeated, poor thing, and expressing feelings of pity and denial is not the way to be in tune with the child's feelings. For example, the child might say, *"I am learning disabled, I cannot do these things, I cannot fill out an application."* Now you are getting down to a concrete problem. Again, the parent mirrors by saying, *"Boy, that sounds like it was really awkward and embarrassing."* This is not pity. You are not judging their feelings but, rather, describing their feelings. Now that you have begun to forge a link, you can continue with, *"In the future, you might be able to bring the application home so that we can fill it out together."* We want to have a working relationship with these kids.

You do not want to take a Pollyanna approach, in which one would comment, *"Everything happens for the best. You did not get that job, but you will get the next one."* We really should not look to rescue children from painful moments. This is not our task as parents. Our task is to help them to integrate the pain, and to help them learn from painful experiences. And the best way they can learn is if they feel you are their ally. They will sense this when they feel that you are attuned to them. To give another example, your daughter might come home and sadly say, *"Bill and I have just broken up."* A response that will not result in a linkage might be, *"You will find another boyfriend, don't worry."* A response that might encourage more involvement would be something simple such as, *"You sound very sad."* This encourages the girl to tell you how she feels, that she really feels down and out, and that she feels that she will not find another boyfriend. At this point in the conversation more constructive problem solving may be possible. Adolescence is a time when daily changes take place. We need to stay close to them and to mirror back their feelings as we see them.

Now, let us consider anger. If 50 or 100 million American parents are getting angry at their kids, then we know it is a normal emotion. It is a part of life. We thus want to teach LD children how to express anger effectively. Many times, they get into trouble because they do not have the tools to do this well. So what we need to do as parents is model anger in effective ways. What we don't want to create is resentment by demanding that they listen to us. They will listen to you and reflect compliance, but then you will get passive aggression on their part, in which they drag their heels and are late for things because they are mad at you.

What we really want to do is teach children how to respond. There are a few basic rules to keep in mind. We want them to describe what they see, to describe what they feel, and to describe what needs to be done. In teaching them this, under no circumstances should you ever attack the child. For example, when the child takes a shower and leaves wet towels on the floor, one might be inclined to go in and say, *"You are so damn lazy that you could not even pick up a towel!"*

It is far more effective to respond by saying, *"When I see towels on the floor, I feel frustrated and upset and I would really appreciate if it you would pick them up."* This is an elemental process. You describe what you see, then say how it makes you feel and, lastly, state what you want done about it. To give another example, when a teenager comes home late, you may be inclined to yell, *"You are so irresponsible, you aren't going out again!"* However, a comment such as, *"When you come home late, I really feel worried,"* is much more helpful. It is always good to encourage a discussion and in this situation you might continue with, *"What do you think we can do about it?"* You do not want to directly attack the youngster. You should simply state that, *"I am really concerned when you come home late because I am worried that something happened to you, and that maybe your car ran out of gas."* Another example is that many kids do not want to do their homework. In this situation, you can get into a major power struggle about their homework. If we engender anger and resentment in a situation like this, our children will always want to get revenge. Helpful criticism involves saying what needs to be done in a situation. You address the situation or behavior, you do not attack the person. For example, if the kid is not studying, an unhelpful comment would be, *"You will never amount to anything in life."* A helpful descriptive comment such as, *"You seem to have a hard time settling down to do this homework,"* is far more encouraging. You have not criticized, you have merely described. Wait to see how the child responds. He might say, *"Yeah, I hate it."* Again, think of yourself as a mirror and say, *"Yeah, I see you really dislike it. It looks really frustrating. What do you think we can do about it?"* If the youngster begins to make a battle out of it, use words such as, *"You are really angry about the homework, but I do not want to have a struggle and I do not want to be on your back. Let's see how we can solve it."*

QUESTIONS AND ANSWERS

Question: If my child has a socially defeating experience which I have also observed, and he does not bring it up, should I be the one to bring it up for discussion?

Dr. Goodman: We must be careful not to be intrusive to children. And we want them to feel that they have some control over their own feelings. However, we may use a type of non-intrusive mirroring as a way of finding out what happened. You might say, "I noted a while ago that you got in an argument and that some of the kids were really mean to you." Something that is descriptive and simple is a good way to start. Children do not want to feel that you are intruding; they need an invitation. They won't deny that some thing happened and they may say, "I hate those kids." You might be tempted to respond with, "Yes, everyone gets into fights sometimes, and that is the way life is." This, however, may shut down communication with the child, for it says you do not feel their pain. However, if you take a different approach such as, "Boy, you are really angry," this is simply reflecting back his feelings. Now you might get to hear more. Many times kids are worried that you will want to talk about things that they are not ready to think about, or that you are going to give them a lecture. But if you keep a low-key approach it may prove encouraging.

Question: What do you do when your child takes anger and frustration out of their siblings?

Dr. Goodman: LD children are often abusers, but we want to teach them how to defend themselves without becoming oppressors. Nor do we want them to become oppressed. We do not want them to be cruel to other people, but we do not want them to be vulnerable to the cruelty of others. Many times a frustrated child, however, turns into a mean child. When we see this anger, we may want to use this mirroring approach and to articulate what their inner experience is. We might want to say, "Joe, you look so angry." We do not want to say, "Joe, when you get angry you do stupid things." Rather, we want to say, "Joe, for you to do that, you must have been really, really angry." You are trying to get in tune with his feelings. After acknowledging that their behavior is unacceptable and that they are capable of better, we can begin problem solving.

Question: What about struggles around bedtime? Even after a shower he is not ready to go to bed until 9:30 or 10:00 p.m.

Dr. Goodman: Your question contained different kinds of feelings. You should use those feelings when you talk to your child and articulate your feelings. You should say, "I do not want to treat you like a little kid because you are really responsible in some areas, but when it is time to go to bed I notice it is really hard for you to get to bed on time." What is good about this approach is that it is a description, not an accusation. This is less likely to provoke defense, and just reflects your feelings. You are saying that you do not want to be restrictive, but you notice he cannot take care of this bed problem. After you have this non-blaming link, you encourage the child to help you think of solutions and what to do about the problem. You are in a non-blaming mode, and with a teenager you can start to solve the problem. If the child continues to express anger such as, "Get off my back," use the mirror once again and he will begin to feel understood. Then you can start problem-solving. Sometimes they may come up with ideas that are even more strict than those you might have come with on your own!

If the child continues to reject the idea of problem-solving, you may have to offer more mature thoughts. For example, better ideas on how to get to bed on time. If you notice your own anger building up when the child does not respond, you should continue to describe your own feelings about anger and frustration regarding the present conflict. Don't let your anger grow very much, and tell the child that you are beginning to get angry and that you are tired; that you do not want to yell and you ask the child for help in taking care of this problem. If this appeal to reason does not work out, you have to discuss alternative plans with your child, like starting to prepare for bed earlier in this case. Thus, if you institute punishment, it is not to inflict pain on the child, but it comes as a logical consequence to not solving the problem of going to bed. You should tell the child if he does not want you on his back that he should help you in solving the bedtime problem.

LECTURE 7

Bryant J. Cratty, Ed.D.
University of California, Los Angeles
Professor Emeritus of Kinesiology

COORDINATION PROBLEMS AMONG LEARNING DISABLED CHILDREN: MEANINGS AND IMPLICATIONS

"Some say that dyslexia doesn't exist, or that if it does, it goes away. But dyslexia spans a range from mild to severe learning disorders, with no single remedy for everyone."

Introduction by Richard L. Goldman

Tonight's presentation will focus on coordination problems of learning disabled children.

Dr. Bryant Cratty is acclaimed for his work with physical education for the disabled child. He is a Professor Emeritus from University of California Los Angeles (UCLA) in Kinesiology. Dr. Cratty has written over seventy books and monographs dealing with motor learning, adapted physical education for the handicapped and developmental psychology.

We are extremely fortunate to have this professional conduct a presentation in our Speaker Series. Ladies and gentlemen, Dr. Bryant Cratty

Dr. Bryant Cratty

Awkward children, whether otherwise learning disabled or not, are likely to elicit various reactions from those observing their struggles as they attempt to perform physical tasks. For example, they may be the victims of social derision from their brothers and sisters. Peers in school may also ridicule them, and ostracize incompetent performers from games and sports. Their teachers may become impatient as printed or written lessons are not copied neatly and rapidly enough to match the efforts of their physically more capable classmates.

Parents, when confronting evidence of incoordination in their offsprings, also have reactions that vary from household to household. Some accurately assess the nature of the problem and seek valid and thorough evaluations and then obtain professional help.[1]

Others may ignore the difficulties their children have when attempting physical skills. While still others may deny there is a problem present, often citing evidence of their child's intellectual social and/or creative prowess as ways to counter criticisms of their offspring's poor motor abilities. Fathers may become tense while attempting to play ball with their poorly coordinated sons in the backyard, and later may withdraw from such potentially useful encounters. A mother, after receiving reports from a nursery school teacher that include the words *"delayed development,"* may frantically begin to seek a solution to the problem by consulting the family pediatrician, and a number of other medical, educational and paramedical specialists.

IDENTIFICATION AND DEFINITIONS

Identification

Physically awkward children may pose evaluation and labeling problems to professionals assessing them, similar to the problems encountered when attempting to identify youngsters with learning disabilities. Some pediatricians, who may have only been exposed to children with obvious symptoms of cerebral palsy, may shrug off symptoms of incoordination as simply a collection of unusual neurological *'soft signs'*. Furthermore some may tell parents that the child may outgrow poor coordination, and thus the problem is not of importance. Many medical specialists, however, are becoming aware of the existence of the significant percentage of these children found within apparently normal populations. These doctors actively assist parents to seek remedial help for their youngster's problems.

Child psychologists, whose evaluation tools usually include collecting samples of printing and figure drawing, may correctly identify a graphic problem. However psychologists usually do not assess deficiencies reflecting the child's integration of the larger muscle groups. Some learning disability specialists, and educational therapists may mistakenly assume that because the child prints slowly and poorly, other academic difficulties must inevitably be present.

Definitions

Awkward child have been given various labels throughout the past several decades. They have sometimes been dubbed *'minimally brain damaged'*. This type of definition infers that poor motor coordination is caused by easily identifiable neurological underpinnings.

Some have referred to physically awkward children as hyperactive, or possessing an *"attentional deficit."* Those formulating this type of label apparently ignore the premise that it is possible for a child to be hyperactive and well-coordinated, or to be uncoordinated and yet evidence good attention and impulse control.

Others have described awkward children as evidencing developmental apraxia. The prefix *"developmental"* usually means that the condition is apparent early in life, and that the awkwardness observed is reflected by the inability to integrate task's sub-movements into complex wholes. Furthermore *"developmental apraxia"* implies that youngsters given this label are not mastering appropriate physical skills at the ages expected by members of their sub-culture.

An early definition of incoordination was formulated by Gubbay (1975), who wrote that *"the clumsy child is to be regarded as one who is mentally normal, without bodily deformity, and whose physical strength, sensation and coordination are virtually normal by the standards of routine conventional neurological assessment, but whose ability to perform skilled purposive movement is impaired."* Gubbay's description poses several problems, however. For example, his definition suggested that the nervous systems does not somehow mediate awkward motor behavior, as he pointed out that conventional neurological assessment will likely reflect normality when applied to such children. However, twenty years ago Touwne and Prechtl (1970) two Dutch neurologists, and more recently others (Tupper 1987), have incorporated measures of motor awkwardness and neuromotor immaturity into well-designed evaluations for children with *'minor nervous dysfunctions.'* Several contemporary neurological-evaluation schedules have been designed to be used following conventional neurological assessments. Thus it appears that using contemporary assessment tools and sophisticated strategies, signs of motor awkwardness are medically identifiable and diagnosable with reference to the presence of associated neurological dysfunctions.

Recently, clumsiness in children has been defined by Hall (1988) as *"a deficit in the acquisition of skills requiring fluent coordinated movement, not explicable by general retardation or demonstrable neurological disease."* This definition seems based upon contemporary evidence making it increasing clear that a child may evidence motor awkwardness independent of, and without the presence of, other medical conditions and learning problems. Furthermore research data appearing in the 1990's is beginning to identify the existence of various sub-syndromes within what was once viewed as a rather global and undifferentiated type of movement problem. (Hoare & Larkin 1991) (Miyahara 1992).

Awkward children and youth, however, continue to reside in a diagnostically *'grey area'*, similar to the manner in which some learning disabled youngsters are still viewed. They make their peers tense and resistive during play, and also present management and educational problems to parents, teachers and to professionals who deal with them. Similar to what is said about learning disabled children in general, awkward children possess hidden disabilities, problems that may only become apparent when complicated motor tasks are expected of them. Thus awkwardness among learning disabled populations presents still another problem with which many learning disabled youngsters, and those formulating programs for their benefit, must deal.

INCIDENCE

Poorly coordinated children are present in significant numbers in groups of youngsters from which the obviously physically, mentally and sensorially handicapped have been identified and removed for special help. The incidence of children evidencing poor coordination, according to various estimates, ranges from 5% (Breamer & Gillman 1966) (Gubbay 1975) (Henderson & Hall 1982) (Iloeje 1987), to 15% (Hoare & Larkin 1990, and even to 20% of school populations, from which the obviously handicapped have been removed (Paine 1968) (Clements 1966). Their presence has been documented in Europe (Gubbay 1976) Asia, Africa, (Iloeje 1987), Australia (Hoare & Larkin 1991), and in North America. Thus throughout the world it appears as though there is at least one awkward child within most classroom settings containing at least 20 youngsters). Thus motor awkwardness appears to be a pervasive problem among contemporary children and youth.

Gender Differences

It has been assumed for several decades that clumsiness is more prevalent among boys than among girls (Henderson & Stott 1977) (Lazlo et al. 1988). However recent evidence suggests that this problem may be as common among young females as among males (Short & Crawford 1984) (Hoare & Larkin 1991). In the past a preponderance of awkward boys, have been usually found among both experimental and clinical populations studied. However, this sex bias may have been due to the tendency to refer boys, more often than girls, to remedial programs.

Motor Awkwardness Among the Learning Disabled

Contemporary data indicate that the incidence of motor disabilities within dyslexic populations is usually greater than is expected within academically average groups. Brying & Michelsson found that 18% of a learning disabled group they surveyed evidenced motor problems (1984). Bruininks discovered a similar percentage of poorly coordinated youngsters among a group of learning disabled children (1978). Klasen places the percentage even higher, finding that 49.2% of the dyslexic group she surveyed evidenced visual-motor difficulties (1972). Jaklewicz after measuring a population of Polish children with dyslexia, over a ten-year period, stated that disorders of motor functions coexist with dyslexia in a rather constant manner. Her findings indicated that not only are motor problems pervasive in dyslexic populations, but also that coordination problems persist in individual youngsters over an extensive period of time (Jaklewicz 1980).

In a study by Sudgen and Wann (1978) it was also found that higher incidence of motor problems existed within a population of children with *"moderate learning difficulties"* than is expected among children without learning problems. They discovered that from 29–33% of the learning disabled children they tested evidenced coordination difficulties, as contrasted to the 5% of children with motor problems who are expected within academically normal populations (Sudgen and Wann 1987). In a recent cluster analysis by Miyahara (1992) it was found that among a learning disabled population numbering one hundred and sixteen boys and girls, over 60% evidenced either several motor problems or a specific coordination difficulty.

Qualitative differences have also been found when contrasting the motor characteristics of learning disabled, dyslexic youngsters to populations who are academically intact. For example, in a series of studies, Peter Wolff and his colleagues identified deficits in timing tasks (finger tapping and the like) among academically deficient populations. (Wolff et al. 1984). They attributed these differences to neurological problems among the learning disabled including either impaired hemispheric communication, or functional difficulties within the left hemisphere (Wolff et al. 1990). They assumed that these differences reflected timing problems common both to reading and to executing relatively simple, rhythmic motor tasks.

HISTORY & BACKGROUND

For centuries dramatic and obvious movement anomalies, observed in both the brain damaged and among the emotionally disturbed, attracted the attention of medical doctors and of scholars within emerging scientific-medical disciplines. Aurelous von Hohenheim, who lived from 1493 to 1541, was among the first within the Western World to describe the signs and symptoms of a number of movement disorders. Included among his writings are descriptions of actions he labeled *'choreas'* (*from the Latin "to dance"*) consisting involuntary spasms seen in the face, trunk, and limbs (Barbeau 1982). For the next four centuries other medical pioneers, with increasing sophistication, explored the causes, symptoms and potential cures of other movement disorders.

However not until the middle of this century has motor incoordination among maturing children attracted much scientific interest. Scientific interest in what was sometimes termed the 'clumsy child syndrome' has been varied over recent decades. The pioneering efforts of Nicholas Oseretsky in Russia during the 1920's signaled the start of this type of focus. The qualities of clumsiness evaluated subsequently were often dependent upon performance demands and upon values attached to physical prowess in various sub-cultures, and upon the orientation of the professionals assessing motor inaptitude in children.

Since the 1960's articles focusing upon the awkward child have appeared with increased frequency in journals of pediatrics, neurology, education and learning disabilities. In 1975 three books focusing upon the awkward children were published. Two had similar titles "The Clumsy Child" (Gubbay 1975) (Arnheim 1975). A third book titled "Remedial Motor Activity for Children" (Cratty 1975) also dealt with the identification and remediation of poor coordination in children. Contemporary work has illuminated the nature of neurological soft-signs, included both sensory and academic symptoms prevalent among children with poor coordination (Tupper 1987). A text dealing with Clumsy Child Syndromes by Cratty (1995) also presents a comprehensive view of the problems of physical awkwardness occurring in developing children and youth.

A contemporary emphasis upon the early recognition of physical awkwardness in children as a medical problem has been evidenced by the inclusion of this condition in standard neurological texts in the 1960's (Walton 1966) (Brain and Walton 1969) (Ford 1966). Often this problem has been categorized as a type of minimal cerebral dysfunction, or as a type of neurological soft-sign, reflected in symptoms of motor incoordination (Deuel & Robinson 1987).

Poor Coordination Among the Learning Disabled

Scholars have interpreted the significance of motor difficulties among groups of learning disabled and dyslexic populations throughout the years. On one hand, Critchley & Critchley (1978) proposed that that developmental dyslexia is due to cognitive disabilities which are *"frequently of constitutional origin"* (page 7). Moreover these same authors suggest that the incidence of motor disabilities within dyslexic populations is *"entirely coincidental,"* and thus (poor coordination) is *'apart from dyslexia'* (page 89). They concluded that *"dyslexic boys are proficient at ball games, while girls may be able in needle work"* (Critchley & Critchley 1978, p 92).

Numerous writers, in the early 1950's began to promise that the identification and remediation of poor coordination in learning disabled youngsters will dramatically improve both school readiness and academic success. The titles of their books and articles often reflected these cheery speculations, and included "How to Develop Your Child's Intelligence," (Getman 1952) and others (Kephart 1955). As a result, some unsuspecting teachers, parents, and other professionals eagerly exposed children to structured regimens of motor activity in attempts to improve reading and other academic, perceptual and cognitive skills. Most of the time, however, the expectations and hopes these theories generated have not been realized. However, the results of a critical survey by Kavale and Mattson (1983) suggested that, (1) not only did well publicized programs of motor remediation not significantly modify learning efforts, including reading, but that (2) better classroom progress would have been obtained through traditional applications of academic tutoring and that (3) the motor programs applied did nothing to improve measures of motor coordination. Other reviews of the literature have also concluded that simply trying to remediate dyslexia through the application of motor program will not succeed in modifying reading or other academic qualities. (Critchley & Critchley 1978) (Myers & Hammill 1989). Reviews of the currently popular sensory-integration approach to remediating academic-reading disabilities by both Lerer (1985) and Gottlieb (1987) also point to the short-sightedness of attempting to remediate academic difficulties with simply-applied sensory-motor strategies.

A Contemporary Viewpoint

Recent research should encourage professional workers to view physical awkwardness with more precision, than was true in past (Miyahara 1992) (Hoare & Lark in 1991). These data are also beginning to encourage the formulation of symptom-specific remedial procedures. Some of the pertinent literature appearing during the past two decades, has begun also to focus upon both poor physical coordination as a problem, separate and independent of academic learning disabilities. Thus the use of motor interventions to modify a laundry list of academic difficulties, including reading is becoming increasingly naive. Lastly newly developed imagining techniques have begun to result in the identification of some of the neural substrates underlying subtle coordination difficulties (Knucley et al. 1983).

The Present Focus and Contemporary Trends

This lecture is an effort to illuminate the causes, characteristics and educational ramifications of learning disabled youngsters with below-average physical coordination. In the United States, during the 1980 and 1990's, poorly coordinated children have become increasingly recognized as an important sub-division of youngsters requiring special testing and services. This recognition is reflected in legal-educational descriptions which guide the diagnosis of all atypical children. The 1987 revision of the Assessment Code, contained in the Diagnostic and Statistical manual employed by psychologists and psychiatrists in the United States, contains detailed descriptions of what are labeled, *"Specific Developmental Disorders"* (Axis 111). One is termed a *"Developmental Expressive Writing Disorder"* (315.80), and a second a *"Developmental Coordination Disorder"* (315.40) reflecting poor integration of the larger muscle groups.

It is believed also that the specialists in learning disabilities, and those formulating curricula for the learning disabled, have important moral, professional and legal obligations to assess and to attempt to rectify motor coordination problems frequently exhibited by the members of the population they serve.

CAUSES AND NEUROLOGICAL IMPLICATIONS

One cause of poor coordination is some degree of oxygen deprivation occurring at birth. All structures of the body are susceptible to hypoxic injury prior to, or during, birth. However, most vulnerable are structures within the central nervous system. Problems encountered by the premature infant, including short episodes during which breathing terminates, results in hypoxic reactions, damage to the respiratory centers themselves, and other functional problems such as a compressed or entangled umbilical cord. Severe may cause death.

When oxygen deprivation to the brain occurs it is likely to influence the type of damage incurred and also the developmental delays later encountered. For example, if oxygen deprivation occurs early during gestation, the deeper layers of the brain including the several centers, controlling movement storage and programming, are likely to be influenced. If apoxia is encountered later in development, damage is likely to be seen in the cerebral cortex, and thus associational-thought processes are likely to be disrupted. However, cerebral damage within each circumstance may be widespread and diffuse, or result in abnormalities related to a more limited area (Towbin 1970).

Vascular accidents prior to, during or shortly after birth also cause motor difficulties ranging from the mild to the severe. These incidents can include rupturing of blood vessels early in fetal formation and thus involve the deeper layers of the central nervous system. In contrast, those occurring later during gestation of shortly after birth are likely to influence more superficial layers of the central nervous system. Near birth the cerebral cortex acquires it venous system, and thus is susceptible to damage due to one or more vascular accidents (mild infarctions).

As is true in the case of hypoxia, vascular incidents are not of all-or none nature. Vascular accidents and incidents occur to varying degrees, as is true of oxygen deprivation. The resultant neural damage due to infractions may thus vary markedly. Marked and obvious disruption of the motor systems occurs when these two problems occur in severe proportions, while minor to moderate symptoms of clumsiness are encountered when these problems occur to mild degrees (Towbin 1987). A sharp blow to the head is one cause of behavioral abnormalities which reflects trauma to the central nervous system. If no penetration of an object occurs these are usually labeled *"closed brain injuries."* Delineating the precise manner in which behavior will be disturbed is not always possible upon knowing the location of the cranium which has received the blow. Striking the head may thus have unpredictable outcomes. This uncertainty of outcomes occurs because a blow can cause a shearing effect, in which surface neurons are scraped in ways that may cause problems difficult to predict. Often swelling produced by a blow in one portion of the brain may cause pressures in other parts, also producing diffuse and/or unpredictable outcomes. Unilateral involvement occurs if the child has been struck on the side of the brain. Movement problems in this case appear on the side, opposite to that receiving the blow. Usually the younger the child or youngster incurring head trauma, the more likely post-trauma motor functions may be positively modified with remedial efforts. A relatively small percent of learning disabled children, with minor, to moderate coordination problems, have suffered some kind of head trauma.

Types of biochemical insults may also cause neural damage reflected in poor motor coordination. One type of problem encountered in large numbers of infants in the 1980's were the influences of drug ingestion by pregnant mothers. These noxious substances include crack cocaine, discussed in the section which follows. Other abusive substances ingested by pregnant mothers including alcohol, and nicotine can cause developmental problems including motor incoordination (Deren 1986). Lead poisoning, as well as exposure to asbestos may cause problems that are reflected partly in motor incoordination and mental confusions in executing recreational skills and self-care tasks.

The limited information that is available indicates that familial influences may also result in awkwardness in some children. In a study, involving more than 53,000 women, sponsored by the National Institute of Neurological Diseases and Communicative Disorders, conducted between 1959 and 1965, developmental parameters including motor coordination were contrasted to variables that included family membership. It was found that the probability of a child evidencing signs of neurological dysfunctions, including motor incoordination, were 30% greater when other peers also were delayed motorically. The familial relationships extended also to linkages of moderate neurological dysfunctions among cousins.

In the same study when the presence of motor soft-signs in forty twin pairs were contrasted, the degree of predictability rose even higher (Nichols 1987). Regehr and Kaplan (1988) also have found that motor inaptitude when coupled with learning disabilities are apparently traceable through families.

Numerous other types of insults to, and problems within the central nervous systems may cause motor incoordination of either an obvious or subtle nature. These include tumors of various types that are often the result of little-understood inherited qualities. Tumors within the central nervous system are of numerous kinds. Their type, size and location will influence development in unpredictable ways. A tumor may cause an organ to malfunction, and/or crowd neural structures thus causing problems by their bulky presence. Sometimes surgical and chemical efforts to reduce tumor's effects results in unwanted symptoms including sensory, cognitive and/or motor deficits.

SUB-TYPES

It is apparent that several sub-syndromes occur within the groups of youngsters evidencing *"The Clumsy Child Syndrome."* It is purposed within this section to identify and to describe the characteristics of the more prevalent sub-syndromes among populations of awkward youngsters among the learning disabled. This identification is believed important for several reasons. The description of viable sub-syndromes should result in more useful diagnostic and remedial guidelines being formulated, than are presently available. Secondly, this analytical undertaking might serve to stimulate needed research focusing upon awkward children among the learning disabled.

The following sub-syndromes were formulated by consulting two types of information. The first source consisted of research studies. In these, investigators employed factor and cluster analyses of performance attributes, obtained from groups of awkward children (Miyahara 1992) (Hoard & Larkin 1991). Groupings of neurological soft-signs have also been explored using factor analytic methods, and have provided another source of objective data (Turner 1987).

Information was also derived from our own clinical observations and upon the experiences of other professionals working with neurologically suspect infants, children and adolescents within remedial educational settings.

CEREBELLAR SYNDROME

Cerebellar functions in humans have been studied for decades. For example, symptoms reflecting a moderately dysfunctional cerebellum have been measured in late childhood and have been associated with low birth weight by Lesny and others (1980). Cerebellar problems are thus likely among numerous populations of developmentally delayed youngsters.

One of the main jobs of the cerebellum is to refine and smooth motor programs, as they emerge from the central nervous system, and journey to the muscle-collections that combine to produce smooth and purposeful action patterns. The cerebellum also is believed by some as responsible for controlling and preprogramming various rapid automatic patterns in which some muscle groups engage. These automated movements may include the rapid throws (saccades) made by the eyes when reading. Other times the eye jumps in this rapid manner when tracking fast moving objects (moving more than about 50 angular degrees per second).

The groups cerebellar-motor symptoms seen in a child also may include tremors and inaccuracies of the larger muscle groups, as well as hand-eye coordination difficulties. Problems controlling precise eye movements are important when engaging in sports containing rapidly moving objects and people, may also be caused by cerebellar dysfunctions.

Larger parts of the cerebellum are devoted to controlling the precise movements of the mouth, lips, and hands, as is true of the main sensory-motor cortex. In contrast, smaller portions of the cerebellum mediate the less precise movements of which the trunk, and lower limbs are capable. This is an important principle to keep in mind, when considering the specific nature of various cerebellar soft-signs that awkward child manifest.

Among the main behavioral signs of some kind of cerebellar *"involvement"* are unsteadiness and mild tremors. As only a portion of the cerebellum may be involved, in some youngsters the tremors may be seen as regular low amplitude fluctuations only in the hands, or of the pencil when drawing, and may not be apparent in the lower limbs. In other children, unsteadiness is only evidenced in gait, or when attempting to balance. The presence of unsteadiness and tremors in both the upper and lower parts of the body is most likely, however. Unsteadiness in the upper limbs may be seen as the child produces lines in which there are marked by tremors, and unsteadiness while walking. Other problems in hand-eye coordination may include dysmetrias, difficulties in making accurate lines, and making reaching movements of exact lengths. Additional problems can include a decomposition of rhythmic and fluid movements, into disjointed and disconnected ones.

Other types of cerebellar symptoms include poor kinesthetic awareness of limb location. This child, for example, may need to watch his or her arm as it is drawn back in an overhand throwing motion, or it may simply *'disappear'* from the thrower's consciousness.

Scientific verification of the existence of a sub-syndrome of cerebellar types' among groups of awkward youngsters may be found within the date emerging from several studies of neurologically suspect children. For example, an investigation by Ritter and his colleagues (1970), it was found that hopping tests frequently identified children with other neurological soft signs present, including problems when walking and turning. Nichols' comprehensive factor analysis, conducted from 1959–1965, using data obtained from several thousand children (1987) also reflected this syndrome. Nichols found that the inability to hop consecutively at the age of 4 years, together with imprecise drawing efforts were both predictive of the appearance of other symptoms later in life including hyperactivity in school settings. More directly to the point, associated signs of clumsiness including measures of poor coordination and abnormal gait. At the same time these two measures 'loaded' in the same factor as signs of synkinesia (mirror movements) and other abnormal movements (Nichols 1987).

Two recent cluster analyses conducted with populations of awkward children further confirm the existence of this sub-syndrome. Miyahara (1992) identified a sub-group within his data, that included with poor dynamic and static balance, coupled with deficits in ball-aiming skills. The linkages described earlier in this section between visual accuracy and balance, both attributes hypothetically mediated by the cerebellum, would account for this association between deficits in ball skills and balance within Miyahara data. Hoare and Larkin (1991) also identified a sub-group of awkward children evidencing specific deficiencies in balance, and the absence of other pronounced motor coordination problems. This group constituted over 28% of the awkward children assessed in their cluster analysis (Hoare & Larkin 1991).

Hypotonic Syndrome

Children beset with the hypotonic syndrome, display levels of resting muscle tone that are below average. This flaccid state may also occur in the facial region, and in other parts of the body. For example a hypotonic youngster may display a 'a sleepy look' as overly relaxed muscles lifting eye lids, result in the eyes remaining half-closed even during waking hours. Among children with this condition the stomach muscles may appear flaccid (ptosis) resulting in a *"pot-bellied"* look. The term benign hypotonia has been used to describe this type of syndrome, indicating that the condition is not progressive (Eng et al 1979).

One of the five sub-types identified by Miyahara's cluster (1992) evidenced hypotonicity and poor fitness, coupled by *"fair coordination."* This was the fifth most prevalent sub-type within the population of learning disabled awkward children measured by Miyahara, and accounted for almost 4% of the youngsters surveyed.

This condition is usually encountered first in infancy, as delays in the assumption of various important stable and static positions are attempted. During the past two years I have assessed and devised developmental programs for three hypotonic infants ranging in age from 18–22 months who could not sit up unassisted. An additional two youngsters, whom we have evaluated, were almost three years of age, and yet had not pulled themselves to a standing position and thus had not begun to walk. The overall physical health of these youngsters was adequate and normal, nor did the condition seem to be progressive in nature. They appeared similar to floppy rag dolls when attempting to sit unaided, to assume a creeping position, or when trying to stand. Home-based programs of infant stimulation were formulated, in which their parents could participate.

The status of most young children evidencing below-average levels of residual muscle tone, remains constant and benign, and they do not develop more serious and progressive neuromotor problems later in life. However in relatively rare cases hypotonia observed in a preschool youngster may be a sign of progressive condition. In the factor analysis by Nichols (1987), both the motor symptoms and personality characteristics of the hypotonic youngsters are revealed. In Nichols' investigation, a measure of hypoactivity loaded in the same factor as did tests denoting *"withdrawal'* and *"social-emotional immaturity"* (Nichols 1987). Nichols' data confirms a compensation frequently seen among hypotonic children, *"learned helplessness."* The hypotonic youngster of ten evidences attentional deficits, but does not frequently engage in hyperactive behaviors. This type of *'dreamy'* youngster often seems out of contact with teachers, parents and peers. He or she may display periodical lapses of both visual and cognitive inattention. Thus the hypotonic child is likely to withdraw both cognitively and socially from both interaction in play and in academic environments (Nichols 1987).

Tension Syndrome

Some youngsters labeled *'clumsy'* display constant and excessively high levels of muscle tone. In contrast to hypotonic children previously discussed, a youngster evidencing a tension syndrome displays signs of hypertonicity while attempting skills, as well as when at rest. The tension levels are raised when a skill becomes difficult for a child to execute. The accompanying tension thus may compound struggles to execute complex skills. This syndrome may also appear as a form of developmental dysgraphia. Furthermore, symptoms reflecting the tension syndrome, within groups of awkward children, may represent mild to moderate forms of spastic cerebral palsy.

This syndrome is often accompanied by excesses in overflow or associated movements (synkinesia). When a hypertonic child is required to engage in a forceful movement with the hands, the mouth may grimace excessively and when vigorous jumping movements are executed, the upper limbs may move and flex. Reaching movements are restrained, because as they attempt to extend their arms toward objects, the smooth elongation of the flexer muscles of the arm do not occur. Instead the biceps may become tense and inhibit coordinated extension. Basic reflexes may be exaggerated in strength when elicited (Nichols 1987).

Children exhibiting excess amounts of muscle tension typically evidence hyperactivity. This includes exhibiting locomotor and manipulative behaviors that are in excess of those desirable within a school setting. Data from the study by Hertzig and Shapiro (1987) suggest the consistent presence of hypertonic signs, within populations of premature children.

Unlike hypotonic children, who often seem to be in dream-like states for a good part of the school day, awkward youngsters evidencing a tension syndrome frequently leave their chairs in a classroom, grab for objects and swipe at other children who may pass in reach. The data from the factor analytic study of neurological soft signs by Hichols (1987) contains correlations between emotional fluctuations, short-attention span, and muscular tensions coupled with impulsivity. Thus there is statistical evidence that confirms the existence of the tension syndrome with accompanying attentional deficits and impulsivity.

Attempting to acquire printing and writing skills usually presents problems for this type of youngster. Learning to print letters may result in the frequent *'overshooting'* of a line, or ending point (dysmetria). This problem is sometimes 'solved' by the child trying to both start a movement and to inhibit a movement(s) at the same time. Thus a great deal of excess tension may accompany efforts to improve graphic skills. Relaxation training is often a helpful accompaniment to writing and printing lessons, when working with this type of child.

Within a clinical setting, children evidencing *'tension syndromes'* need careful and patient attention. Frequent rests from tension producing practice may be necessary. Breaks during which relaxation training, or similar techniques may be applied, are often useful. At the termination of motor skill lessons it may be necessary to apply additional relaxation strategies to reduce their tension and to produce a positive emotional state before they proceed home, or to other parts of a school program.

Dyspraxia Syndrome

Some awkward children have difficulties effectively planning and chaining together series of sub-movements into smooth, well-integrated motor skills. Their levels of muscle tone may not be excessively higher or lower than average, nor do they evidence the unsteadiness reflecting some kind of cerebellar involvement. Their problem may be reflected primarily in poorly sequencing sub-actions of movements into whole skills. They may be identified as dyspraxic.

Miyahara identified a sub-group among the awkward learning disabled children he surveyed who evidenced poor "dynamic coordination." This was the third most prevalent syndrome within his population consisting of over 30% of the total group surveyed, (Miyahara 1992). Two Australian researchers also found a group within the population they measured evidencing poor motor abilities independent of perceptually *'loaded'* tasks. For example, these apparently apraxic children did poorly in tasks involving complicated assembly work (Hoare & Larkin

1991). These youngsters constituted about 10% of the group surveyed by these researchers *"down-under"* (Hoare & Larkin 1991).

After formulating a movement idea, a child afflicted with the apraxic-dyspraxic syndrome may often act impulsively, rather than displaying the self-control typical of neurologically more intact youngsters. These portions of the frontal-associational centers may be also be deficit in these youngsters. Hypothetically portions of the central nervous system of a severely dyspraxic youngster has received an insult, either hypoxia or a vascular accident, relatively early in prenatal development (Towbin 1987). This kind of neural pathology thus may have influenced the adequate functioning of one, or more, of the several neural structures believed to inter-act in their control of motor programming, storage and retrieval.

Symptoms of dyspraxia may also appear more markedly within some parts of the body than within others. Deficits in motor planning skills may, in various children, lie primarily in the hands, limbs, or when they attempt to move their total body with accuracy through space. I recently tested an eight year old boy who executed drawings resembling those of a talented adult artist. However, when asked to replicate movements of his limbs and total body he proved completely confused. The presence of these specific types of limb and trunk apraxia, resulted in this boy evidencing an extremely low self-esteem. He appeared severely depressed because of the social abuse of peers consistently accompanied his efforts to participate in playground activities. However, several months of individualized instruction resulted in measurable improvement in his ability to engage in sequenced playground tasks.

Dyspraxia in these children may also be reflected in speech-articulation difficulties. Correctly sequencing the complex movements of the tongue, lips, and other parts of the mouth, when forming word sounds, may present profound problems for youngsters evidencing the *'apraxic-dyspraxic syndrome.'* Numerous neurological schedules assess speech-articulation problems along with other movement-type soft signs (Ritter 1970) (Hertiz & Shapiro 1987). Speech pathologists often remark at the unusual percent of children with gross motor problems who come to them with speech difficulties. This problem has also been labeled oro-apraxia, and buccofacial apraxia.

When working clinically with these type of children, it is important initially to expose them to various tests of motor-planning. It has been found that motor planning deficits, and strengths, are task specific (Cratty 1989). When found deficits in these types of tests these children and adolescents should be taught multi-faceted skills using progressive-part methods. For example, a complex skill should be first broken down into parts, and after a single component is introduced and mastered, it should be combined with a second, then a third part, until the entire skill is mastered. Exposure of these children to impatient teaching and to ill-prepared instructors, who demand the quick mastery of skills, are likely to result in negative emotional outcomes. Gordon & McKinaly have also written about strategies useful when working with dyspraxic youngsters (1986).

Manual Graphic Syndromes

It is common to evaluate children whose main symptoms only reflect difficulties with tasks requiring manual dexterity and/or who display problems when attempting to draw letters and numbers. Two separate global manual syndromes appear to exist. On one hand there are children whose fingers fumble when attempting to button their clothes and who often drop their forks and spoons while eating. Some of these, while holding their pencils in a fist-like grip, will display average or above average abilities to reproduce figures, letters and numbers.

In contrast, there are other children whose dexterity is adequate, but whose graphic skills, including letter reproduction ranges from labored to nearly absent. Indeed the factor analytic literature strongly suggests that in adults there may be several more independent manual-dexterity qualities or abilities within their overall *'movement profiles.'* (Cratty 1975). Dysgraphia is frequently seen among the learning disabled. In a learning disabled population we recently surveyed, this problem was found to occur in from 30–40% of those assessed (Miyahara, Cratty, and Goldman 1990). Dysgraphia, however, may also be unaccompanied by other academic problems. For example, in a recent investigation of children with writing and printing difficulties, it was found that about one-half of the youngsters surveyed evidenced dysgraphia without the presence of other learning disabilities (O'Hare & Brown 1989). Dysgraphia, unaccompanied by other academic difficulties, has itself been labeled a *'specific learning disability'* by Brown (1981) and others (Myklebust 1973).

Among the contemporary classification systems proposed is that by O'Hare and Brown (1989). They divide dysgraphia into two main categories. One is labeled specific dysgraphia, and includes three motor coordination sub-groups, and a fourth reflecting spelling problems. A second main classification they called non-specific dysgraphia. This classification, they suggest, includes writing problems caused by cognitive and experimental variables, difficulties typically seen within mentally handicapped and culturally deprived populations. In general, among dysgraphic children one is able to identify specific sub-syndromes, roughly corresponding to those listed within this section, including dyspraxics, hypotonic and hypertonic youngsters, as well as unsteady tremorous ataxic printing and writing.

VISUAL–PERCEPTUAL CLUMSINESS

There are numerous visual and visual perceptual problems that may result in awkward motor behaviors. These problems may range from ocular-motor problems involving the erratic or inefficient movements of the muscles moving the eyes, to severe retinal deterioration. In addition, the camera-like eye may take a clear and accurate *'picture'*, but the child may not interpret, and organize visual information well and quickly. This latter type of confusion is often referred to as perceptual rather than a visual difficulty.

Hoare & Larkin (1991) recently identified a sub-type within the population of clumsy children they surveyed as evidencing problems in tasks heavily loaded with visual–perceptual components. This sub-group, consisting of 18% of the group they surveyed, had difficulty in *"visually loaded"* tasks while evidencing average and above-average performance in other motor skills.

Visual and visual perceptual problems interfere with accurate movement because the eyes and motor systems are neurologically and thus functionally intertwined in innumerable ways. Poor balance may occur, for example, because the eyes are unable to focus well thus destabilizing the child's space field. This type of problem may be seen when the youngster is able to balance in a static position better with the eyes closed, than when they are open. In these cases a confusing visual field, when temporarily eliminated by closing eyes, permits the motor system mediating balance to function better. In contrast, when some visually and/or perceptually inadequate children attempt balance task with their eyes open, below average scores may result. Problems causing unclear vision may also inhibit accurate movement.

The presence of erratic (nystagmatic) eye movements, that may reflect the presence of various visual pathologies, may also prevent a child from intercepting and extracting information well from moving objects. Serious visual-neurological problems may also result in inability to intercept moving objects well. These can include the presence of a tumor, at any one of several points within the optic pathways in the front of the brain (the optic chiasm). These types of tumors may restrict the size of children's visual fields, and thus prevent them from tracking balls through the same wide arc, available to youngsters whose visual fields are normal in size.

Accommodation problems, also commonly prevent children from focusing upon written work at a desk near the eyes. This problem involves the inability of the ocular muscles to draw the eyes toward each other and to focus at written work on the desk. Thus accommodation problems are likely to inhibit effective hand-eye coordinations needed when learning to print letters. Most of these difficulties, however, are modifiable with corrective lenses.

Children whose visual systems are apparently intact may lack the abilities to organize and to interpret important visual information. They appear to be beset with perceptual rather than with visual difficulties. This quality also has been found to be multi-dimensional. Groups of tests evaluating visual perceptual (organizational and interpretational) skills are applied to large groups of children and factor analyzed, a number of separate and independent qualities emerge (Smith and Smith 1966). These included depth perception, the ability to fragment space (what is halfway between you and the door?) and the ability to judge movement in space. Thus it appears that deficits in innumerable visual and perceptual problems, acting alone or in combinations, are likely to affect physical skills in negative ways.

Mixed Syndrome

In addition to youngsters who evidence clearly definable sub-syndromes, some awkward children display symptoms that make their precise classification difficult. Thus the symptoms some awkward children display may also cause them to be correctly labeled *'mixed'*. For example, it is often difficult to separate tensions and incoordination that may, because by neurological causes, from difficulties arising an emotional overload, be the result of continual task failure caused of poor motor planning abilities.

Ocular control problems may not only cause motor incoordination, but also be part of a pervasive coordination problem arising from central nervous system dysfunctions, also resulting in imprecise movements of the larger muscle groups. Moreover, as Laszlo and her colleagues have pointed out, motor clumsiness is often due to a combination of motor incoordination, integration of process problems, and to deficits in monitoring kinesthetic sensory input (1985, 1988). Such subtle variables as differences in leg length, poor foot mechanics or other postural-structural problems can also influence motor functioning in youngsters.

Youngsters falling within this mixed classification may evidence a variety of coordination problems, reflected in virtually every motor task with which they are confronted. Both Miyahara (1992) and Hoare and Larkin (1991) identified sub-types who evidenced pervasive and consistent difficulties in all the tasks within the test batteries they employed. In Miyahara's investigation this sub-group consisted of 17% of the population he surveyed. In an Australian population measured by Hoare and Larkin (1991) the group of youngsters who were *"consistently below average in all tasks"* comprised 19% of the group evaluated. Moreover, in both investigations even lower-functioning groups appeared in the data.

Overview

Acceptance of the validity of the presence of various syndromes suggests that each poorly coordinated, learning disabled youngster, should be considered as unique. A knowledgeable evaluator should assume that few general qualities and parameters exist, or that predictable correlations will be obtained between tasks sampling various skill groupings. Moreover, the presence of these various sub-types among learning disabled children suggest the importance of instituting syndrome-specific interventions that are carefully and developmentally planned.

SENSORY–PERCEPTUAL PROBLEMS

There are pervasive types of sensory and perceptual-motor difficulties that are commonly found among various sub-categories of awkward children. For example, it appears more likely that a child who is learning disabled and awkward will display a larger number of perceptual deficits than will be presented when an academically average or superior child is evaluated.

Body Agnosias and Dysgonias

The word *'agnosia'* refers to a lack of awareness that may appear in various sensory modalities. The term dysgnosia has also been used to denote disruptions of sensory awareness, (Lesny 1980). Visual agnosi, as thus refer to the poor visual perception of various stimuli, including word-blindness, and reading problems. Auditory agnosias, in contrast, denote auditory perceptual problems.

There are also a number of *'body agnosias'*. These deficits may involve sensory-perceptual difficulties that children have when attempting to locate their bodies in space. Tactile deficits in the limbs and the fingers, and also poor kinesthetic awareness of limb location and movement are also contained within this general group of symptoms.

In the 1980's researchers began to link these sensory deficits occurring in children to neural abnormalities. In one study, for example, it was found that 50% of the youngsters, who evidenced dysgnosias, also had abnormalities in various components of their central nervous system. These structural differences included ventricular enlargement (Knuckey et al. 1983).

The symptoms reflecting body agnosias may be inseparable from behaviors seen in youngsters diagnosed as evidencing dyspraxias of various types. Thus confusions occurring when moving in space, and when sequencing task movements, may be attributed both to the presence of dyspraxia, and also to the presence of dysgnosias. For example, if a child has difficulties dressing himself/herself one observing clinician may identify the problem as a dressing apraxia, while another will state that the problem reflects a body-agnosia (a type of somatoagnosia). Thus the same, or a similar, behavioral deficit is likely to reflect the interaction of both an agnosia and also of a specific type of apraxia.

There are measurable sub-components to various types of body agnosias and dysnosias. For example, digital agnosia, reflecting a lack of sensory-awareness of the fingers and/or the hands, is composed of several sub-components, including (a) deficits in the ability of the fingers to *'record'* sensations and obtain tactile information when objects (placed out of view) are touched or stroked. This quality has been termed "stereognosis or, if deficit, astereognosis (b) the ability of the child to identify fingers when they are named verbally and (c) the ability to determine where tactile stimuli are applied on one hand, when viewing the other hand (Gerstmann 1940). Thus digital agnosia should be considered in specific ways, and with reference to the test employed to evaluate different sub-types.

For over one hundred years descriptions of deficits in the body image may be found in the neurological literature. Initially the only descriptions of agnosias available were those made among stroke patients (Obsersteiner 1881) (Luria 1966). During recent decades, however, agnosias have been studied among maturing youngsters evidencing various developmental sensory and motor, problems. Initially a few, selected sensory deficit were described among awkward children brought to clinics for evaluation. Gradually the measures increased in number. Even now, however, the majority of measures in examinations designed to assess the presence of neurological problems in awkward children, evaluate motor rather than sensory qualities (Tupper 1987, Appendix).

Spatial Agnosias

General spatial agnosias refer to problems youngsters may have when attempting to transport their total bodies through space with accurate sequenced movements. Difficulties are encountered by youngsters evidencing this type of problem, when trying to move around the furniture in a room, or when attempting to pass between classmates and school desks. The bumping that may occur among peers may result in awkward children experiencing social problems.

The remediation of, and accommodation, to this type of problem may involve both communicating an understanding of the problem to the child's caretakers, and helping the child understand his or her problem in clear terms. The presence of these spatial confusions should not be interpreted by parents as reflecting their off-spring's ignorance, or lack of willingness to respond correctly to directions. Rather both the child and parents should recognize the presence of the problem and should accommodate to it in various ways. For example a series of directions, to "go up stairs and to put on a red jacket," should be made shorter and simpler. Most of the time partial directions should be given at first.

Tactile Agnosias of the Body

In addition to deficits in the heurological mechanisms mediating the mental-spatial maps needed to move the total body through space, are problems reflected in the poor tactile awareness of the body's surfaces. Tests to ascertain the sensitivity of the various surfaces of the body often constitute parts of neurological evaluations. For example, a sub-test assessing of two-point discrimination is often employed. This involves determining how closely two points may be placed together and touched to various body surfaces, before they are perceived by the child as a single touchpoint. At times, the clinician-physician may simultaneously stroke both sides of the body in order to determine possible differences in tactile sensitivity. If differences are found they have important diagnostic significance, including the possible presence of tumors, and thus these symptoms should be interpreted by those with appropriate medical backgrounds.

Limb Agnosias

In clinical practices, dealing with awkward children, it is common to encounter young clients who evidence the inability to locate their limbs in static positions. Furthermore they also have difficulties when attempting to kinesthetically monitor the movements of their arms and legs. When throwing a ball, for example, such youngsters may keep their throwing arms in front of their bodies. Poor kinesthetic acuity measured in limb positioning, and in movement tasks among awkward children has been verified in data from studies by Hulme and his colleagues (Hulme et al 1987), by Smythe and Glencross (1986) and by Judith Laszlo and her colleagues (Laszlo et al. 1980) (Laszlo 1985, 1988).

When attempting to remediate this kind of problem, it is often necessary to augment kinesthetic input, with tactile stimulation, and with visual information. For example, when teaching a child to throw, a mirror placed to the side may prove a helpful teaching aid. The throwing arm may thus be viewed as it is drawn behind the thrower's line of vision. If the problem is severe, tactile input may be helpful. The arms may be gently stroked or rubbed in order to afford more information as to their location and volume. Children whose movement problems include cerebral palsy have traditionally been exposed to this type of tactile intervention. Foot placement when attempting various locomotor tasks, such as skipping and hopping, may be assisted if templates are placed on the ground, into which the feet may be placed.

Digital Agnosias

Digital agnosia is the poor awareness of the location and tactile proprieties of the hands and fingers. There has been a long history of interest in this problem, Digital agnosias include several sub-types. These include problems reflected in tasks assessing sensitivity to touches received by the hand. Difficulties in determining the shape, texture and shape when touching or handling objects reflects another type of digital agnosia, sometimes termed astereognosis. It has been found in several studies that awkward children, when compared to normals, exhibit less tactile sensitivity in the hands (Kinnealey 1989), and also receive less accurate information when both touching and handling objects with their hands (Haron & Henderson 1985).

Gerstmann identified digital agnosia as one of three problems, within a triad of symptoms, including difficulties in math (acalculia) and body image deficits (Gerstmann 1927). Difficulties in carrying out simple mathematical operations is associated with digital agnosia, some have suggested, because these youngsters have problems using their fingers as counting implements when gaining initial quantitative concepts (Strauss and Werner 1938).

In a recent study my students and I found significant positive correlations between measures of hand sequencing and mathematics abilities, confirming the possibility of a sub-group of children evidencing Gerstmann's syndrome, within learning disabled population evaluated (Miyahara, Cratty and Goldman 1990). In other studies by Lord & Hulme (1987) and by Hulme and his colleagues (1982) it has been found that clumsy children as a group are likely to be measurably deficient in their ability to obtain accurate tactile-kinesthetic information about the lengths and shapes of various objects when they are inspected manually.

Benton (1959) and others have formulated direct measures of finger, or digital agnosia. Essentially these usually consist of asking a child to identify which fingers are touched or stroked by an examiner, on a hand that is kept out of view of the child. At times the identification of the finger touched is made verbally and at other times the child is asked to move the finger, or fingers, involved. In the second case, a movement quality probably influences the scores obtained.

In real life situations, a child afflicted with this kind of finger insensitivity may experience a number of difficulties. I recently tested a six-year old girl, who while scoring high on tests of intelligence, was eating with her hands, rather than with table utensils, due to the presence of a severe case of digital agnosia. Often children with digital agnosias, use a fisted grip on their pencils while writing, they gain so little tactile information from their pencils when they are held in the usual way. Attaching four fingers to the pencil thus afford them more tactile-sensory information, than what is gained if the pencil or pen is be held in the traditional way.

Children with signs of various types of digital agnosia often avoid making small models. They also find typing their shoes difficult or impossible. They may fumble when attempting to use zippers, or buttons on their clothing. Catching balls can prove difficult as they obtain late and inaccurate information from the missiles as their hands contact incoming balls. They may also be fearful of climbing playground apparatus as they are unable to feel the bars in their hands when attempting to grasp them.

EMOTIONAL ACCOMPANIMENTS

Those studying and chronicling the characteristics and habits of awkward children have been virtually unanimous in the observation that various emotional problems accompany physical inaptitude. In many ways these are similar to the problems usually found among the learning disabled, and reflects compensations for feelings of low self-esteem. It seems logical to assume that negative social feedback and poor self-assessments resulting from the inability to write well in a classroom, coupled with poor play skills, will likely result in lowering both awkward children's and learning disabled youngster's self-esteems. Available evidence from contemporary, data-based studies is beginning to confirm this hypothesis.

Data Based Evidence

In one type of study, comparisons have been made between the self-concept of children labeled clumsy, in contrast to how physically average children report feeling about themselves. In an investigation of this type conducted by me and post-doctoral students several years ago, a questionnaire-type self-concept test was used containing questions reflecting children's feeling about their appearance and physical ability. The test was originally constructed by Dale Harris and Ellen Piers (1964).

In four of the twenty questions tendered to the 222 male subjects, differences were found between the two populations contrasted (Cratty et al. 1972). The awkward boys reported, more often than did the psychically adequate boys, that they were sad most of the time, that they did not believe themselves strong, and that they would rather watch than play games. A significantly larger percent of the awkward boys also reported that reading was not easy for them. This final finding perhaps reflects the presence of a significant percent of boys with learning difficulties often found within populations of awkward children.

A student and I recently followed up this 1972 investigation with another, using a highly similar questionnaire (Dalrahim & Cratty 1990). Eighty-three awkward boys, averaging 6.3 years of age, were polled relative to the feelings they had about themselves, and how they felt about their physical appearance and ability. These were children, whom we had evaluated during the past thirty-six months at our clinic and averaged delays of from 1 to 1.5 years in physical coordination. The scores of 17 awkward girls were used in the same study, using the same questionnaire. They averaged 6.5 years of age, and also were delayed physically from one to two years. Of the twenty questions there were marked differences in the percent of low-self-concept responses of awkward children obtained in the 1990 study contrasted to the response of normal subjects in the 1972 study (Cratty et al.).

In a second type of investigation, comparisons are made of the motor competencies of children within various atypical groups. Including in these have been groups of children labeled learning disabled, the mildly and moderately retarded, and the emotionally disturbed. An example is a study by D.H. Stott and his colleagues (1972) in which comparisons were made of children evidencing varying degrees of emotional maladjustment based upon their scores from The Bristol Social Adjustment Guide (Stott and Marston 1971). Stott found that the emotionally disturbed population identified, was over represented among awkward children. Hostility scores were found to be closely associated with motor impairment, as were scores reflecting the tendencies to both over-react to the environment, and to under-react and behave in overly passive ways in social situations. Stott found that *'inconsequence'* (failure to inhibit impulses) was the quality most highly associated with motor impairment. The scores Stott obtained from the Bristol Social Adjustment Guide also correlated with mean scores from the tests of motor impairment administered.

Summarizing this work, Stott pointed out that three times as many maladjusted children, compared to well adjusted youngsters, were motorically impaired: And likewise three and a half times as many motorically impaired youngsters, as compared to motorically normal children, were maladjusted (Stott et al. 1976).

The data from other, more contemporary, studies by Hulme and his colleagues (Hulme et al. 1982) (Abbir et al 1978) and Henderson and Hall (1982) also indicate motor ability tests can statistically discriminate between groups of emotionally normal youngsters and groups of youngsters evidencing various psychiatric syndromes. Data from additional investigations, including those by Van Rossum and Vermeer (1990) and Kalvergbor and his colleagues (1990) and others (Henderson et al. 1989) also indicate that there are strong links between physical awkwardness, low self-esteem, and social problems.

Factors analyses have also been conducted containing tests reflecting both emotional qualities and movement qualities, including tests of coordination and motor *'soft signs.'* For example, in a study by Nichols (1987), tests of hyperactivity, emotional stability, and attention span loaded in a factor containing scores from two drawing tests. Thus the overall picture from this and similar studies is that motor inaptitude likely causes various symptoms reflecting emotional upset.

The relationships between emotional instability and motor impairment may be circular. Stott, for example, suggested that the interactive relationships he found were due to the presence of general types of neurological impairments that may have influenced both motor and emotional qualities in the youngsters he evaluated. The correlations found in the studies that have been reviewed, thus require close scrutiny in order to determine possible causal relationships present, and the direction of the causation.

Data from a few studies are now beginning to illuminate the possibility of modifying emotional states by exposing awkward children to remedial programs. In recent work by Laszlo and her colleagues, for example, direct evidence was obtained of 'dramatic' changes in social–emotional behaviors occurring among children exposed to a program, that also produced parallel and positive changes in their physical skills (Laszlo et al. 1988).

Most encouraging, however, are the recent appearance of longitudinal studies surveying, not only motor characteristics, but also the emotional states of youngsters assessed during several testing sessions taking place throughout the formative years of life. These studies are obtaining evidence obtained during significant time periods within the lives of the young subjects studied. In one of the first of these it was found that, while some awkward children improve over time, many continue to have difficulties, of several types, well into their teens (Gillberg & Gillberg 1989).

A comprehensive, longitudinal study was published recently by Anne Losse and her colleagues. They compared measures of emotional health and physical coordination, obtained from thirty-two youngsters at both their sixth and sixteenth years of life. They found strong evidence that the youngsters surveyed continued to have motor problems, attracting the attention to those attempting to teach them motor skills during this entire decade of their lives. During the several testing sessions necessary to complete this study, it was also observed that many displayed *"intense personal feelings of failure"* throughout this ten year period.

It was concluded by Losse and her helpers, that clumsiness is not a benign disorder confined to childhood, but rather that awkwardness continues into the teens. Their data also made it clear that persistent and measurable evidence of motor ineptitude during the years the study took place, was accompanied by feelings of social inadequacy among most of the youngsters taking part in the study. However they also observed that among some subjects who had been extended continuous and effective parental support, symptoms of poor emotional health were significantly reduced, and in some cases apparently eliminated (Losse et al. 1991).

The available data from both correlative and longitudinal investigations thus strongly suggest that motor ineptitude both causes, and is accompanied by, low self-esteem and associated social adjustment problems in many children and adolescents for significant time periods during their formative years. Motor task performances constitute concrete, and observable, measures of competency, out of incompetency. A youngster who cannot perform well thus sees vivid evidence of his or her failings, and concurrently receives negative social feedback from observing peers, parents, and teachers. Thus both self-perceptions and the social judgements of others combine to lower the self-esteem of many awkward children and adolescents.

Clinical Observations of Compensations, and Avoidance Strategies

A number of compensatory behaviors reflecting inadequate social-emotional adjustment are likely to be exhibited by awkward children. Parents, professions in clinics, and teachers are among those whose best efforts are often thwarted by these, often abrasive, ways in which awkward children act. These behaviors are expressed, and the forms they take are both important to consider, when assessing and providing programs for children with coordination problems.

Engaging in these strategies serve several important needs on the part of a clumsy youngster. (1) A compensation sometimes reflects the child's attempts to totally avoid situations in which physical performance will be required, (2) in another group of behaviors that are seen within performance situations the child has not been successful in avoiding. Thus while on the playground the child finds ways of actually not participating. (3) Still a third type involves ways clumsy children have of modifying both their own performance and performance demands so that they have a chance to succeed, but on their own terms.

Often an awkward child will engage in several of these avoidances–compensations, in quick succession, or at different times throughout a single day. A compensation or compensations may be selectively used when an awkward child is confronted with specific demands made by adults, children and by various situations encountered. That is why many youngsters will learn to carefully match a specific compensation to various demands and stresses. It is typically found that each child favors one or more of these compensations, and consistently relies upon them. Some of the more common of these compensations are as follows.

School Phobias

Some poorly coordinated children may attempt to avoid school attendance completely. They may do this for various reasons, including the knowledge that they must take a fitness test in their physical education class that day. Upon arising in the morning their effort to avoid school may range from claiming illness, to *'throwing'* temper tantrums. Their protestations may even be accompanied by measurable medical symptoms. They may become nauseous when contemplating their school day, and accompany these feelings with pleas to remain at home.

Learned Helplessness, and Infantilization

One of the most pervasive, and potentially harmful, compensations involves the display of learned helplessness. Infantilization may be partly the reflection of inappropriate caretaker behaviors, including the tendency of some mothers and fathers to unknowingly reward "babyish" behaviors. Infantile behaviors arise from the fact that developmentally delayed children often are subtly rewarded for their incompetence and inabilities to perform physical tasks well. The problem can also stem from an attitude, adopted by the clumsy child, that most of what he or she attempts will likely be unsuccessful. Thus they give up trying when confronted with physical performance situations. If not encouraged by caretakers to try to accomplish various motor task, over time helpless feelings become more pronounced.

Symptoms within this general pattern of helplessness include displaying immature play patterns, and a disinclination to attempt or to persist at motor tasks. These symptoms may even include delays of speech and language. Infantile speech patterns are common among children who have learned to be physically helpless. Signs of social and emotional immaturity displayed by an infantilized child may include frequent temper tantrums, thumb sucking and similar behaviors that are not age appropriate.

This pattern of behavior, if continually reinforced, will persist into adolescence. Parents who continue to refuse to demand mature behavior and performance will likely produce psychologically and physically helpless teenagers. It is striking to observe these symptoms evidenced by some adolescent learning disabled adolescents, for example.

The answer to this problem is usually intensive family therapy, so that the mother and other caretakers are provided a mirror in which they may view the ways they are rewarding, and thus maintaining, their offspring's helpless and ineffectual behaviors. After gaining an awareness of how their parenting practices influence their children, attempts would then be made to modify these behaviors. However, often *'selling'* the need for such therapy is a task that professionals contacting parents are not always able to accomplish.

Comedic Behaviors and Other Types of Compensatory Verbalization

Within play situations, the awkward child runs the risk of not been noticed. Being overlooked may appear to such children to be a form of punishment worse than receiving ridicule. Thus the uncoordinated youngster often makes an effort to make the other children know they exist. One strategy is to use an excess of verbal behaviors at play, and in other situations. These verbal behaviors may include (1) A constant attempt to be funny, by engaging in excess comedic behaviors (2) Compensatory bravado, consisting of statements that express an overly self-confident demeanor, *"I can make the Olympic Team!!"*, (3) Constantly voicing statements that have nothing to do with a play situation, and (4) verbal preoccupation with the rules, acting like a *'playground attorney.'*

The content of hyperverbalizations varies. This excess chatter may consist of persistent attempts to joke and be funny, including making fun of the play situation and of other children. At other times this hyperverbal behavior consists of intellectualizations, about the skills required or about other aspects of the games. Still another focus of this chatter may be upon rules and their interpretation. In this latter case the awkward child may sound like a youthful attorney, while engaging in sophisticated arguments about scores, rules and the like. This type of verbal behavior thus may also effectively block requirements to exhibit physical skills in the games confronting the youngster.

Aggression

Lacking adequate play skills, awkward children sometimes engage in excess physical and/or psychological aggression against others. Sometimes this aggression appears in the form of verbal abuse, or even of obesities directed toward peers. At other times aggression may consist of simply chasing weaker or younger children around the playground. At still other times direct physical aggression is engaged in. Case studies of young adults who have evidenced marked sociopath behaviors often contain scenarios of how, as children, they lacked adequate play skills and thus began to take out their frustrations in the form of aggressive vengeance against others.

MOTOR COORDINATION PROBLEMS AMONG THE LEARNING DISABLED

Within the past decade, several studies have been conducted focusing specifically upon the motor abilities, and their meanings, among learning disabled children. For example, Denckla and her colleagues (1985), explored possible relationships between dyslexia, attention and motor proficiency, and concluded that there were various sub-groups within learning disabled populations, including those with and without attentional and motor problems. In this same study it was also found that the tests of motor qualities they employed, including measures of synkinesia (overflow), and of rapid-repetitive rhythmic movements, were more likely to predict hyperactivity in children than were measures of specific learning disabilities they obtained. Most important, their data suggested to them that treatment outcomes may be different for the various sub-groups. Outcomes seemed to depend upon the presence of various combinations of learning disabilities, hyperactivity and of types of motor incoordination measured within each sub-division.

In two other factor analyses by Nichols (1978) and by Ayres (1972) it was found that motor coordination and sensory-qualities *'loaded'* in factors separate from those containing motor coordination items. It is interesting to note however, that Ayres then proceeded to formulate and promote a program of sensory-integration in which motor-sensory experiences were advanced as producing change in a variety of academic and sensory-perceptual processes. Moreover Ayres work has been recently criticized for the capricious manner in which factors were labeled within the various studies she carried out from 1965 and 1987 (Cummins 1992).

The most reasonable contemporary viewpoint thus seems to be that motor development programs should be one, of several parts, of curricula for learning disabled youngsters. However, simply exposing a youngster with learning disabilities to sensory-motor experiences will do little to modify academic competencies (Kavale 1987) (Gordon & McKinlay 1989) (Lerer 1983).

The data from investigations conducted at Landmark West from 1989–1992 have provided useful information about the motor abilities and disabilities of learning disabled children (Cratty, Miyrahara & Goldman 1991) (Myrahara & Cratty 1991). As a group the youngsters in these studies displayed motor coordination qualities inferior to current developmental averages.

These investigations also produced several findings useful to those planning school programs for this type of population. For example, it appears that learning disabled girls may present special problems when confronted with physical activity and physical education. They may lack both motor activity capacities, and also harbor extremely negative feelings about how they both look and function in motor skill situations. The data revealed that the females were even further behind norms, than were their male classmates, when measures of coordination, fitness-strength, and self-concept were averaged. It appears that special attention would thus be paid to special needs of learning disabled girls when programming for physical improvement. The data we and others have obtained related strongly suggest that many learning disabled youngsters are not happy with their body's appearance and with the sports skills and fitness they are able to exhibit.

The data also suggest that important sub-groups are present both within populations of awkward children and also within groups of the learning disabled (Lubs et al. 1991). On the basis of this data it appears that learning disabled children may be divided into at least four sub-groups, relative to physical competencies. These include one group which is relatively free of motor problems, and indeed may be able to compensate for academic deficiencies through vigorous and sustained participation in recreational sports, and in related activities.

A second sub-group seems to consist of individuals who evidence motor planning problems along with academic difficulties. There is recent evidence that this sub-type may be pervasive within family complexes (Regehr & Kaplan 1988). Rie also refers to the problems learning disabled children have executing complex serial skills (1987). A third group may evidence adequate skills and fitness, but require a longer period of time to acquire complex skills because of the presence of dyspraxias of various kinds.

Still another sub-group of the learning disabled may be able to learn skills reasonably well, but lack the muscular and cardiovascular fitness and the coordination needed to fully and vigorously participate in sports and games. This fourth group seems to possess physical capacities that remain underdeveloped, because they apparently lack the motivation to participate in vigorous physical activities.

Important to consider is the lack of age-trends in the motor planning scores obtained in these studies. This finding suggests that basic problems in sequencing may persist in learning disabled populations throughout life. If this information is confirmed in subsequent investigations, older learning disabled children and adolescents should also be patiently taught new and complex skills, as is true when dealing with younger children within this type of population. Longitudinal studies, also substantiate the persistence of motor problems, and neuro-motor soft signs, among groups of clumsy children for as long as ten years (Losse et al 1987).

Overall this and similar data suggest that motor coordination problems are an additional difficulty with which many learning disabled children must contend. Thus helping the learning disabled to improve physically is an important goal to be pursued along with strategies intended to elicit academic competencies.

EDUCATIONAL STRATEGIES FOR IMPLEMENTING CHANGE

A question frequently asked by both parents and teachers who observe youngsters evidencing clumsiness is, *"Can anything really be done to change a child's coordination?"* Professionals formulating programs for awkward children should be prepared for these questions during parent conferences following evaluations. They should attempt, furthermore, to offer the best answers possible, by referring to the best available data and to expert opinion. Several variables should be taken into consideration when formulating an answer. These include the age and emotional state of the child, the degree of awkwardness present, and the skills in which improvement is needed.

A comprehensive study used *"neurologically impaired"* children as a sub-population. It was completed in 1968 by G. Lawrence Rarick and his then doctoral student Geoffrey Broadhead (Rarick & Broadhead 1968). Their study explored the possible influence of several variables upon change in a number of physical abilities. Two major subject populations were employed including a group of 206 *"minimally brain damaged"* boys and girls, placed within two age groups, one a younger group of six to nine year olds, and a second group ranging from ten to thirteen years of age. A second group consisted of 275 *"Educable Mentally Retarded"* children were divided into the same two age groups.

Several variations of instruction were applied in this study, thirty-five minute daily, five days a week, for a twenty week period. These included one group taught whose members were physical education individually, and a second group instructed as a small class. Two control groups were also present. One of these remained in their regular physical education classes. The second control group was given art instruction in order to control for the possible effects of special attention upon the changes recorded in the two experimental populations.

The results of Rarick's study were highly promising. For example it was found that significant changes in most of the motor performance tests were recorded by the experimental group exposed to an individualized program. The children labeled minimally brain damaged (who might be labeled hyperactive-learning disabled today) recorded significantly greater changes in motor abilities than did the retarded youngsters. The older male children with minimal brain damage improved more in overall motor abilities, than did the girls, or groups of younger children.

In 1972, my colleagues and I completed a study in which we contrasted the possible improvement in motor abilities measured in a group of twenty-eight awkward boys and girls (7–12 years of age) with whom were working at UCLA. Their scores were contrasted to motor abilities measured in a control group of learning disabled children from a local school district. The control group engaged in a regular program of physical education, while the experimental group came to the University twice weekly for one-hour programs of individualized work in motor skills and coordination. The program included work on drawing-printing skills and also contained the teaching of various gross motor skills. The children in both groups were evaluated initially, and at the end of a five month period. The scores of the trained subjects, were matched and contrasted by age and sex to those of the controls (Cratty et al. 1972).

The data also offered some support for the efficacy of a special program of motor development. Overall the percent of improvement was over 2.7 times greater by the experimental group, than was true among the control group when the average scores, reflecting percent of increase, from the battery of gross motor tests were contrasted. Additionally the mean scores from a figure drawing task also evidenced more improvement among the experimental subjects (14.4%), than was recorded in the data from the control group (1.1%).

Additionally the progress in various types of gross motor tasks revealed that the most marked improvement through training, occurred in balance, locomotor agility, and the catching of balls. Less improvement was noted in scores denoting the identification of body parts, and in gross agility (ascending quickly from a back lying position, and copying a four-part movement of the total body). Encouraging was the finding that the ability to execute complicated geometric figures, adding one part at a time, was highly modifiable through training.

The data was also analyzed to determine within what age ranges the percentage of improvement was most apparent. As can be seen, the improvement among the fourteen younger children from 7–9 years, was more marked than among the fourteen children whose ages ranged from 10–12 years of age.

There are indications in the data that it is not inevitable that possible changes in physical coordination will occur through maturation. The controls, even though exposed to a regular physical education program, regressed in some ability areas. Similar regression in physical attributes, as a function of disuse, has been found in a more recent fitness study among the learning disabled (Cratty and Delrahim 1992).

Within recent years, both clinicians and researchers have suggested strategies for remediating dyspraxia symptoms in children (Gschwend 1988) (Gubbay 1975). Seaman and DePauw, for example, listed twenty-one activities purporting to remediate fifteen signs of dyspraxia in children. However their suggestions were not accompanied by research data substantiating their worth (Seaman and DePauw 1982).

Gschwend (1988) also purposed a program for dealing with dyspraxia in children. This program included music and art therapy, playing chess, and other types of cognitive training labeled *"will training."* Gschwend's suggestions thus appear to constitute a more global approach to generalized cognitive-motor strategies than was true in the more skill-specific program suggested by Seaman and DePauw (1982).

In a recently completed study, it was significant improvement in upper-body strength that was elicited by exposing learning disabled children to a fitness program (Delrahim & Cratty 1991). In this same study, marked improvement in strength was also found among sub-groups of learning disabled girls, and pre-adolescent boys and girls. Most interesting, however, was evidenced that control populations who did not participate in training recorded decrements in fitness and strength, measured over time periods lasting from six to twelve months in duration.

Longitudinal studies which began to appear in the late 1980's corroborated the persistence of coordination problems, along with social-emotional difficulties, among groups of motorically deficient youngsters (Gillberg & Gillberg 1989) (Losse et al. 1991). In the study by Losse (1991) for example, it was found that motor problems, reflected in the presence of motor soft signs, were present not only at the age of six when her thirty-two subjects were first evaluated, but were also seen in evaluations conducted ten years later. While some of the children surveyed over time had made progress, the majority had experienced stress from teachers, and also negative social pressure from peers during their childhood and adolescent years. Gillman and Gillman (1989) have also found that while some of their subjects had improved over time, many had continued to evidence coordination difficulties.

The data from these investigations, and from earlier studies, suggest that remediation of motor difficulties may take place, but only with the institution of effective, individualized programs, that are carefully staffed and planned and developmentally precise. Moreover the information obtained in comprehensive effort by Losse and her colleagues in London, suggests that only with sustained and effective parental support are poorly coordinated youngsters likely to weather the social and academic *'storms'* the are likely to encounter during their childhood and adolescent years (Losse 1991).

There have been several experimental studies conducted, whose authors attempted to determine if the reduction of motor soft-signs, including indices of poor coordination, occurs after exposing children to mediation intended to improve attention span and reduce hyperactivity. The findings from these studies have been mixed. Skekim and his colleagues conducted a study to determine whether soft-signs were reduced as the results of medication (3-methoxy-4-hydroxy-phenethyleneglycol, norepinephrine metabolite). The study was not well controlled, however. The reduction of soft signs noted may have been due not only to the effects of medication, but also because the subjects matured during the relatively long time interval during which study was conducted (Skekim et al. 1979).

Lerer and Lerer (1976) studied the effects of methylphenidate upon forty hyperactive children who exhibited three or more soft-signs. Sixteen of the twenty children in the experimental group evidenced a reduction of soft-signs, in contrast to no seductions among children in the twenty-member control group. However the study suffered from several methodological problems including the use of neurological evaluation of questionable objectivity, and the lack of adequate controls.

In still another study of the effects of medication, McMahan and Greenberg (1977), found that some of the subjects, in both the experimental and control groups, evidenced a reduction in soft-signs. However, in both experimental and control groups several of the forty-four hyperactive children in the study evidenced more soft signs at the conclusion of the experiment!

Wherry and Reeves (1987), and McMahan and Greenberg (1977), have highlighted the problems when trying to ascertain if medical interventions will reduce the presence of soft-signs, including indices of motor incoordination. The vexations they wrote about also included the lack of reliable measures of soft-signs, and the sometimes paradoxical effects of medications when applied to hyperactive children.

Further conundrums are seen in the data from a study by Ross (1982). They found that the medication they employed resulted in the childrens' teachers reporting them more manageable and educable. However, significant changes in academic scores were not forthcoming. This kind of finding has also surfaced when learning disabled children are medicated. These latter data point to the probability that, while medication usually improves attention and reduces hyperactivity, medication needs to be paired with intelligently applied remedial strategies and within well conceived program, in order for significant improvements in both academic and motor abilities to occur. Ross (1982) also expresses concerns about the use of medication in an extensive review of the topic. Both clinical observations and experimental data thus support the judicious use of medication in programs for hyperactive, poorly coordinated children, but only if paired with intellectually applied remediation.

Wherry and his colleagues also noted, in a medication change study, that improved response time in learning tasks occurred in children who were given chemical help to control their hyperactivity (Wherry et al. 1969). It is hoped that research in the future will further explore the efficacy of potentially useful medical accompaniments to traditional forms of motor therapy.

In summary, the somewhat limited evidence available suggests that, under optimum circumstances, some motor skills of young children are modifiable within limits. The more changes seem to occur in younger rather than in older populations, and in certain types of motor tasks in contrast to others. There are apparently optimum ages at which to apply a structured-instructional program in various motor skills. Between the ages of about four years and six or seven years seems best.

Changes recorded in investigations in which in ball throwing for distance and accuracy has been assessed, indicate that these qualities in normal, and among awkward young children, may be moderately improvable through training. In contrast, locomotor activities (hopping, skipping and the like) appear to be more resistant to change through education in groups of young children. Available information also indicates that drawing and printing tasks also seem modifiable through the application of effective educational strategies that include specific motor practice of the tasks involved (Cratty & De Oliveira 1989) (Cratty et al. 1972) (Chapter 7).

In the few available studies of awkward children, the data indicates that coordination tasks, and related activities seem modifiable with training. Moreover, remedial efforts may be assisted with the judicious application of medication when hyperactivity accompanies motor incoordination. However, the presence of soft signs reflecting poor coordination may persist through childhood and into the teens. With effective interventions, the changes that are recorded may occur in specific and useful *"splinter skills."* However, the underlying neurological deficits seem to remain. Finally, interventions throughout an awkward youngster's life need to be accompanied by supportive behaviors emanating from family members, in order to reduce or eliminate undesirable social-emotional side-effects (Losse 1991).

Also, the available data indicates that retrogression in fitness and physical coordination may take place if effective remediation is not instituted (Cratty et al 1972) (Cratty & Dalrahim 1992). The best progress appears to take place when programs are carefully designed, and individual differences are provided for (Rarick & Broadhead 1968).

As is true of other remedial programs, including physical therapy, speech-language enrichment and occupational therapy, the outcomes of attempts at improving the motor coordination of an individual child are often difficult to predict. Moreover, the nature of the possible neural adjustments, underlying modifications of physical skills, are also not well known at this time.

SUMMARY AND OVERVIEW

The data reviewed indicates that many learning disabled youngsters display motor coordination qualities inferior to current developmental averages. These investigations have also produced several findings useful to consider when planning school programs for this type of population. For example, it appears that learning disabled girls may present special problems when confronted with physical activity and physical education. They may lack both motor activity capacities, and also harbor extremely negative feelings about how they both look and function in motor skill situations.

The marked delays in fitness recorded in the studies surveyed could be explained in several ways. One viable supposition is that poorly coordinated learning disabled youngsters tend to avoid physical activity in general and thus evidence the results of this avoidance in scores reflecting poor fitness levels (reflecting what has been termed a *'disuse'* syndrome). When working with groups of learning disabled children, close attention should be paid to motivational factors and to employing strategies designed to enhance their self-esteem. Most of the data we and other have obtained strongly suggest that many learning disabled youngsters are not happy with their body's appearance and with the sports skills they are able to exhibit.

Current evidence also indicates that important sub-groups are present both within populations of awkward children and also groups of the learning disabled (Lubs et al. 1991). On the basis of this data, it appears that learning disabled children may be divided into at least four sub-groups, relative to physical competencies. These include one group who is relatively free of motor problems, and indeed may be able to compensate for academic deficiencies through vigorous and sustained participation in recreational sports, and in related activities. A second sub-group apparently consists of individuals who evidence motor planning problems along with academic difficulties.

There is recent evidence that this sub-type may be pervasive within family complexes (Regehr & Kaplan 1988). A third group may evidence adequate skills and fitness, but require a longer period of time to acquire complex skills because of the presence of dyspraxias of various kinds.

Still another sub-group of the learning disabled may be able to learn skills reasonably well, but lack the muscular and cardiovascular fitness and the coordination needed to fully and vigorously participate in sports and games. This fourth group seems to possess physical capacities that remain underdeveloped, because they apparently lack the motivation to participate in vigorous physical activities.

Data from several studies indicate that motor planning problems in the learning disabled are pervasive. For example, Rie refers to the problems learning disabled children have executing complex serial skills (1987). Important to consider in these data, in addition, is the lack of age-trends in the scores obtained. This finding suggests that basic problems in sequencing may persist in learning disabled populations. If this is true, older learning disabled children and adolescents should also be patiently taught new and complex skills, as is true when dealing with younger children within this type of population.

Further longitudinal studies of motor planning abilities and disabilities among this population are currently underway. Data from longitudinal studies available substantiate the persistence of motor problems, and neuro-motor soft signs, among groups of clumsy children for as long as ten years.

Overall, this and similar data suggest that motor coordination problems are an additional difficulty with which many learning disabled children must contend. Thus helping the learning disabled to improve physically should be an important goal to be pursued along with strategies intended to elicit academic competencies. Not attending to the coordination, fitness, and sport participation needs of learning disabled children, displaying deficiencies in these task areas, is likely to blunt the self-esteem of youngsters who do not need additional evidence of their incompetencies. The elevation of motor-skill and fitness capacities may have some *'overflow'* effect into academic work. However, further research is needed to determine the manner in which the intervening variable, self-esteem, is likely to be positively influenced by the acquisition of increased physical competencies, and then in turn to find out if elevation of the self-concept will positively influence classroom effort and performance.

BIBLIOGRAPHY

Abbir M. Douglas, H., & Ross, K. (1978). The clumsy child syndrome: observation in case studies referred to the gymnasium of the Adelaide Children's Hospital. *Medical Journal of Australia, 1*, 65–69.

Arnheim, Daniel D., & Sinclair, W. A. (1975). *The Clumsy Child: A program of motor therapy*, C. V. Mosby, St. Louis.

Ayres, A. J. (1972). Types of sensory integration dysfunction among disabled learners, *American Journal of Occupational Therapy, 26*, 13–18.

Barbeau, A. (1982). "History of movement disorders and their treatment" in Barbeau A, (Editor). *Disorders of Movement*, Lippencott, Philadelphia, PA. p. 1–29.

Benton, A. L. (1959). *Left-right discrimination and finger localization, development and pathology*, Heober-Harper Books, New York.

Bremmer, M. N., & Gillman, S. (1966). Visuomotor ability in children—a survey, *Developmental Medicine and Child Neurology, 8*, 686–703.

Bruininks, R. H. (1978). *Bruinink's Oseretsky Test of Motor Proficiency*, American Guidance Service, Circle Pines, Minnesota.

Brying G. & Michelsson, K. (1984). Neurological and Neuropsychological deficiencies in dyslexic children with and without attentional disorders. *Developmental Medicine and Child Neurology, 26,* 765–773.

Clements, S. D. (1966). *Minimal brain dysfunction in children.* (USPHS Publication #141, US Government Printing Office. Washington D.C.

Cratty, B. J. (1975). *Remedial Motor Activity for Children,* Lea and Febiger, Philadelphia, PA.

Cratty, B. J. (1989). *Perceptual and Motor Development of Infants and Children, 3rd Edition,* Prentice-Hall Inc, Englewood Cliffs, New Jersey.

Cratty, B. J. (1995). *Clumsy Child Syndromes, Description, Evaluation and Educational Implications.* Harwood Academic Press, Philadelphia Penna.

Cratty, B. J., & DeOliveira, I. J. (1989). *Printing Behavior in Children: Seven Case Studies,* Unpublished Monograph, Department of Kinesiology, University of California, Los Angeles.

Cratty, B. J., Delrahim, M. (1991). The Self-concept of children with coordination problems, Accepted Brazilian Journal of Movement Studies.

Cratty, B. J., Miyahara, M., & DeOliveira, I. J. (1991). Wahrnehmungs-und motorische Fahigkeiten von lernbehindereten Kindern Und Jugendlichen (Perceptual and Motor Abilities of Learning Disabled Children and Youth). *Motorik, 14,* 173–184 (in German).

Cratty, B. J., Ikeda, N., Martin, M. M., Jennett, C., & Morris, M. (1972a). Changes in selected perceptual-motor attributes of children with moderate coordination problems. In Cratty, B. J., Ikeda, N., Martin, M. M., Jennett, C., & Morris, M., *Movement Activities, Motor Ability and the Education of Children,* Charles C. Thomas, Springfield, Illinois.

Critchley, M. & Critchley, E. A. (1978). *Dyslexia defined.* Charles C. Thomas. Springfield Illinois.

Cummins, R. A. (1991). Sensory-Integration and learning disabilities: Ayres' factor analyses reappraised. *Journal of Learning Disabilities, 24,* 160–168.

Delrahim, M., Cratty, B. J., Goldman, R. (1991). The effects of training on the power-strength development of learning disabled youngsters. Accepted, Brazilian Journal of Movement Studies.

Denckla, M. B. (1973). Development of speed in repetitive and successive finger-movements in normal children. *Developmental Medicine and Child Neurology, 15,* 635–645.

Deren, S. (1986). Children of Substance Abuse: A Review of the Literature, *Journal of Substance Abuse Treatment, 3,* 77–94.

Gerstmann (1927). Fingeragnosie und isolierte agraphie; ein neues syndrom, *Ztschr. Neurol. Psychiat., 108,* 152–177.

Getman, G. N. (1952). *How to develop your child's intelligence, A research publication,* Luverne, Minnesota, G. N. Getman.

Gottlieb, M. I. (1987). Chapter 31, Attention Deficits Disorders, Hyperkinesis, and Learning Disabilities, Controversial Therapies in Gordon N & McKinlay I. (Eds). *Neurologically Handicapped Children: Treatment and Management,* Blackwell Scientific Publications Boston pp 251–261.

Gschwend, V. G. (1988). Neuropsysiologisch Grundlagen der Bewegungstherapien. (Neurophysiological bases of movement therapy). Der Kinderazt, 19, 1589–1597.

Gubbay S. S. (1975). *The clumsy child: A study of developmental apraxic, and agnosic ataxia,* W. B. Sanders, Co, Philadelphia.

Hall, D. M. B. (1988). Clumsy Children, *British Medical Journal, 296,* 375–376.

Henderson, S. E. & Hall, D. (1982). Concomitants of clumsiness in young school children, *Developmental Medicine and Child Neurology, 24,* 448–460.

Henderson, S. E., & Stott, D. H. (1977). Finding the clumsy child: Genesis of a test of motor impairment. *Journal of Human Movement Studies, 3,* 38–48.

Hertiz, M. E., & Shapiro, T. (1987). Chapter 4. The assessment of Nonfocal Neurological Signs in School-Aged Children, in Tupper, E. E. (Editor), *Soft Neurological Signs.* pp 71–94. Grune and Stratton, New York.

Hoare, D. & Larkin, D. (1991). Coordination Problems in Children, No 18 State of the Art Review, National Sports Centre, Australian Sports Commission, Canberra, Australia.

Hoare, D. & Larkin, D. (1991). Kinesthetic abilities of clumsy children, *Developmental Medicine and Child Neurology, 33*(8). 871–678.

Hulme, C., Biggerstaff, A., Morgan, G., & McKinlay, I. (1984). Visual, kinesthetic and cross-modal judgements of length by clumsy children: A comparison with young normal children. *Child care; health and development, 10,* 117–124.

Iloeje, S. O., (1987). Developmental Apraxia among Nigerian children in Enugu, Nigeria, *Developmental Medicine and Child Neurology, 29,* 502–507.

Jaklewicz, H. (1980). Follow-up studies on dyslexia and dysorthographia. *Psychiatria Polska, 14,* 613–619.

Kavale, K. & Mattson, P. (1983). One jumped off the balance beam: Metanalysis of perceptual-motor training. *Journal of Learning Disabilities, 16,* 166–173.

Kephart, N. C. (1956). *The slow learner in the classroom,* Charles E. Merrill, Columbus Ohio.

Kimura, D. (1976). The neural basis of gesture qua language, In H., Whitaker & H. A. Whitaker (eds). *Studies in Neuolinguistics,* V1, Academic Press, New York.

Kinnealey, M. (1989). Tactile functions in learning disabled and normal children. Reliability and validity considerations. *Occupational Therapy Journal of Research, 9,* 3–15.

Klasen, E. (1972). The syndrome of specific dyslexia. University Park Press, Maryland.

Knuckey, N. W., Apsimon, T. T., & Gubbay, S. S. (1983). Computerized axial tomography in clumsy children with developmental apraxia, and agnosia, *Brain & Development, 5,* 14–19.

Laszlo, J. I., Bairstow, P. J. (1980). The measurement of Kinesthetic sensitivity in children and adults, *Developmental Medicine and Child Neurology, 22,* 454–464.

Laszlo, J. I., Bairstow, P. J. (1983). Kinesthesis: Its measurement, training, and relationships to motor control *Quarterly Journal of Experimental Psychology, 35,* 411–421.

Laszlo, J. I., Bairstow, P. J. (1985a). Perceptual-motor behaviour: Developmental Assessment and Therapy. Holt, Rinhart and Winston, London.

Laszlo, J. I., Bairstow, P. J. (1985b). Kinesthesis sensitivity test. Rinehart and Winston. London.

Laszlo, J. I., Bairstow, P. J., Bartrip, J. & Rolfe, U. T. (1988). Clumsiness or perceptuo-motor dysfunction? in *Cognition and Action in Skilled Behaviour.* A. M. Colley & J. R., Beech, (editors). Elsevier Publishers, BV, North Holland.

Laszlo, J. I., & Bairstow, P. J. (1986). *Perceptual-Motor Behaviour* Holt, Rinehart & Winston, New York.

Lerer, R. J. (1981). An open letter to an occupational therapist *Journal of Learning Disabilities.* 22–23.

Lerer, R. J., & Lerer, M. P. (1976). The effects of methylphenidate on the soft neurological signs of hyperactive children. *Pediatrics, 57,* 521–525.

Lesny, I. A. (1980). Developmental dysgraphia-dysgnosia as a cause of congenital children's clumsiness, *Brain and Development, 2,* 69–71.

Lord, R., & Hulm, C. (1987). Kinesthetic sensitivity of normal and clumsy children, *Developmental Medicine and Child Neurology, 29,* 720–723.

Losse, A. Henderson, S. E., Elliman, D., Hall, D., Knight, E., & Jongman, M. (1991). Clumsiness in children-Do they grow out of it? A 10-year follow-up study, *Developmental Medicine and Child Neurology, 33,* 55–68.

Lubs, H. A., Duara, R., Levin, B., Jallad, R., Lubs, M. L. Rabin, M., Kushch, A., & Gross-Glenn, K. (1991). Chapter 4 "Dyslexia sub-types: Genetics Behavior and Brain Imaging." in D. D. Duane, & D. B., Gray eds. *The Reading Brain: The Biological Basis of Dyslexia,* New York Press, Parkton Maryland, pp. 89–118.

Luria, A. R. (1966). Higher Cortical Functions in Man, Basic Books, Consultants Bureau, New York.

McMahan, S. A. & Greenberg, L. M. (1977). Serial Neurologic examination of hyperactive children, *Pediatrics, 59,* 384–587.

Miyahara, M. (1992). Sub-types of learning disabled children and youth, based upon measures of gross motor functioning. Unpublished study, Department of Kinesiology, UCLA.

Miyahara, M., Cratty, B. J., & Goldman, R. M. (1990). Praxic behaviors among learning disabled Children and youth, Submitted to *Journal of Movement Behavior.*

Myers, P. I. & Hammill, D. D. (1990). *Learning Disabilities: Basic Concepts, Assessment Practices, and Instructional Strategies.* Pro-Ed, Austin Texas. p. 439–441.

Myklebust, H. R. (1973). *Developmental disorders of written language: studies of normal and exceptional children,* Grune and Stratton, New York.

Nichols, P. L. (1987). Minimal Brain Dysfunction and Soft Signs: The Collaborative Perinatal Study. in Tupper, D. E. (editor). *Soft Neurological Signs* pp. 179–200. Grune & Stratton, New York.

Obersteiner, H. (1881). On allochiria; a peculiar sensory disorder, *Brain, 4,* 153–163.

O'Hare, A. E. Brown, J. K. (1989). Childhood dysgraphia: Part 1 An illustrated clinical classification, *Child: Care, Health and Development, 15,* 79–102.

Oseretsky, N. A. (1931). *Methoden der untersuchung der motorile,* Heft 57. Beihefte Zur Zeitschrift Fur Angewandte Psychologie, Barth, Leipzig.

Paine, R. S. (1968). Syndromes of 'minimal cerebral damage,' *Pediatric Clinics of North America, 15,* 779–801.

Piers, E. V. & Harris, D. B. (1964). Age and other correlates of self-concept in children, *Journal of Educational Psychology, 55,* 91–95.

Rarick, G. L., & Broadhead, G. L. (1968). *The effects of individualized versus group oriented physical education programs on selected parameters of the development of educable mentally retarded, and the minimally brain injured children.* Unpublished Monograph, University of Wisconsin, Madison, Wisconsin.

Reeves, J. C. & Wherry, J. S. (1987). Chapter 10, Soft Signs in Hyperactivity, in Tupper, D.E. (Ed). *Soft Neurological Signs,* Grune & Stratton, New York.

Regehr, S. M. & Kaplan, B. J. (1988). Reading disability with motor problems may be an inherited sub-type. *Pediatrics, 82,* 204–210.

Rie, E. D. (1987). Chapter 9 Soft Signs in Learning Disabilities, in Tupper, D. E. (Ed). *Soft Neurological Signs,* Grune & Stratton, New York. (pp 201–224).

Rie, E. D., Rie, H. E., Steward, S., & Rettemnier, S. (1978). An analysis of neurological soft signs in children with learning problems. *Brain and Language, 6,* 32–46.

Ross, D. M. & Ross, S. A. (1982). *Hyperactivity: Current Issues, Research and Theory,* New York, John Wiley and Sons.

Rutter, M. Graham, P. & Yule, W. (1970). A neuropsychiatric Study in childhood. *Clinics in Developmental Medicine* Nos 35–36, Lippencott, Philadelphia.

Seaman, J. A. & Depauw, K. P. (1982). The New Adapted Physical Education, A Developmental Approach. Mayfield Publishing Co, Palo Alto, California.

Short, H., & Crawford, J. (1984). Last to be chosen the awkward child, *Pivot, 2,* 32–36.

Skekim, W. O., Dekirmenjian, H., & Chapel, J. L. (1979). Urinary MHPG excretion in minimal brain dysfunction and its modification by d-amphetamine. *American Journal of Psychiatry, 136*, 667–671.

Smythe, T. R., & Glencross, D. J. (1986). Information processing deficits in clumsy children. *Australian Journal of Psychology, 38*, 13–22.

Stott, D. H., & Marston, N. C., & Neill, S. J. (1971). *The Bristol Social Adjustment Guides*, 4th Edition, Educational Testing Services, San Diego.

Stott, D. H., Moyes, F. A., & Henderson, S. E. (1984). *Test of Motor Impairment*, Henderson revision, The Psychological Corporation San Antonio, Texas.

Strauss, A. A. & Lehtinen, C. E. (1947). *The Psychopathology and Education of the Brain-injured Child*, Gruen and Stratton, New York.

Sugden, D. & Wann, C. (1987). The assessment of motor impairment in children with moderate learning disabilities. *British Journal of Educational Psychology, 57*, 225–236.

Touwen, B. C. L., and Prechtl, H. F. R. (1970). *The Neurological examination of the child with minor nervous system dysfunction*, London, Clinics in Developmental Medicine, Heinemann.

Tupper, D. E. (1987). *Soft Neurological Signs*. Grune & Stratton, New York.

Van Rossum, J. H. A., & Vermeer, A. (1990). Perceived competence: A validation study in the field of motoric remedial teaching. *International Journal of Disability, Development and Maturation, 37*, 71–81.

Wolff, P. H. & Cohen, C., & Drake, C. (1984). Impaired motor timing control in specific reading retardation, *Neuropsychology, 22*, 587–600.

Wolff, P. H., & Michael, G. F., & Drake, C. (1990). Rate and Timing precision of motor coordination in developmental dyslexia, *Developmental Psychology, 26*, 349–359.

LECTURE 8

Dennis P. Cantwell, M.D.

University of California, Los Angeles
Joseph Campbell Professor of Child Psychiatry
Director of Residency in Child Psychiatry

PHARMACOLOGICAL INTERVENTIONS FOR CHILDREN WITH LEARNING AND PSYCHIATRIC DISORDERS

"Even though I may become frustrated again, I know that I can succeed and will succeed, with maybe a little more work than others."

Introduction by Richard L. Goldman

I'd like to welcome everyone and I am gratified by tonight's attendance. Tonight's topic is extremely relevant, controversial, and misunderstood. To educate us, we are extremely fortunate to have as our speaker, Dennis P. Cantwell, M.D. who is the Joseph Campbell Professor of Child Psychiatry at University of California Los Angeles (UCLA) Neuropsychiatric Institute. Dr. Cantwell was a member of the American Association on Mental Deficiency Task Force on Classification of Mental Retardation as well as on the task forces to develop DSM-III, DSM-III-R and DSM-IV. He is the author of numerous journal articles, book chapters and books such as *Fundamentals of Child and Adolescent Psychopathology*. His accomplishments as a clinician, teacher, reseracher, and administrator have been recognized by his receipt of several prestigious awards, honors and prizes, including the American Psychiatric Association Award for Research in Psychiatry.

To speak about Psychopharmacological Interventions for Children with Learning and Psychiatric Disorders, it is my great pleasure to introduce Dr. Dennis Cantwell.

Dr. Dennis P. Cantwell

This chapter provides an overview of the current thinking about the use of medication treatment for common childhood psychiatric problems. Most of the children seen by us at the UCLA Neuropsychiatric Institute have multiple problems in the areas of learning, cognitive functioning, social skills, interpersonal relationships, and mood alterations. In addition to medications, these youngsters usually require a variety of other interventions such as educational assistance, family counseling, and individual psychotherapeutic help. Generally, the medication is given to ameliorate a certain specific set of symptoms; whereas the other interventions are given to ameliorate a different set of specific problems.

A significant amount of time elapsed between the drafting of this chapter and its publication. During this time new data have appeared in many of the areas discussed in the chapter. Readers interested in an update are referred to:

1. *Textbook of Pharmacotherapy for Child and Adolescent Psychiatric Disorders*. D. Rosenberg, J. Holttum, S. Gershon. Brunner/Mazel, New York, 1994.
2. *Child and Adolescent Clinical Psychopharmacology*, Second Editions. W.H. Green, Williams and Wilkins, 1996.

Our knowledge base about the use of psychotropic medications in children and adolescents has expanded enormously in the last two decades. Nonetheless, compared to our knowledge about psychotropic medications for adults, what is known about children still lags far behind. In fact, this is one area that the Federal Government (National Institute of Mental Health) has targeted as a priority for additional research.

There are several reasons why the use of medication is especially complicated with children, and why less is known about medications for children. First, as noted above, most children have a complex of problems, symptoms, or diagnoses rather than a single problem, symptom, or diagnosis. Multiple diagnoses within one child pose difficulties both for treatment and for evaluating the effects of medication.

In addition to the problem of multiple diagnoses, another problem in evaluating medication effects is symptom differences. Children with the same diagnosis do not necessarily have the same symptoms, although they do have the same types of symptoms or *"symptom clusters."* Examples of symptom clusters include: *"anxiety symptoms," "depressive symptoms," "inattention symptoms," "impulse control symptoms,"* and *"hyperactivity symptoms."* The hyperactivity symptom cluster consists of symptoms such as: fidgeting, squirming, having difficulty remaining seated, excessive running around or climbing on things, moving about excessively during sleep, and always being "on the go." But, all "hyperactive" children do not necessarily exhibit all of the symptoms in the symptom cluster.

A third problem with regard to medication treatment of children is that symptoms and symptom clusters do not necessarily mean the same thing across children. Consider, for example, two boys who both manifest *"hyperactivity symptoms."* One boy manifests *"hyperactivity symptoms"* much of the time, and has the diagnosis of *"attention deficit disorder with hyperactivity."* The second boy has the diagnosis of an *"anxiety disorder,"* and manifests the hyperactivity symptoms only during periods of high anxiety. If given stimulant medication, the first boy should experience a decrease in his activity level. Conversely, if given stimulants, the second boy would likely become more active.

This example illustrates that the same symptoms in different children may reflect different underlying diagnoses. When they do, the children will respond differently to the same medication. As well as showing that there are diagnosis-specific effects of medication, this example also highlights the critical importance of a very careful evaluation and a different diagnosis before any medication is started.

Medication is typically prescribed based on the fact that a child has a particular diagnosis and that, within that diagnosis, a medication is known to produce certain positive symptomatic effects. However, one difficulty with this approach is that medication has two very different types of effects. One is the direct effect of producing certain physiological phenomena that result in the modification of certain symptoms. The other is an indirect or psychosocial effect which, in the long run, can be as important as the direct effect.

For example, consider an 11 year old boy who still wets the bed. For this boy, an anti-neuritic medication (e.g., Tofranil or Imipramine), will have the *direct* effect of blocking urination, probably through a local effect on the bladder. However, the drug will also have the *indirect* effect of allowing the boy to go on sleep-overs or to attend overnight camp without any fear of embarrassment due to bed wetting. To the young patient, this indirect effect is probably more important than the direct effect.

Below, we will outline some of the major childhood and adolescent psychiatric disorders and discuss the pharmacological interventions that are appropriate for them.

ATTENTION DEFICIT (HYPERACTIVITY) DISORDER

The most common childhood psychiatric disorder is *"attention deficit (hyperactivity) disorder"* (ADD). This disorder is characterized by three symptom clusters: hyperactivity symptoms, impulsivity symptoms, and attentional symptoms.

The most important symptom cluster in this disorder is the attentional symptom cluster. Children with ADD have attentional problems such as difficulty zeroing in on the right stimulus (for example, the teacher's voice rather than a neighbor's voice) and difficulty keeping their attention focused on particular stimulus. They are highly distractible, and, consequently, tend to have serious problems in school.

Although these children have an attention deficit, it is not true that they are incapable of attending to anything. In fact, they can attend quite well to certain types of stimuli, for example, Nintendo and action television (i.e., Teenage Mutant Ninja Turtles or World Wrestling Federation matches). They can also learn some things quite easily (e.g., sports statistics), but they usually have difficulty learning reading, mathematics and other types of higher-level cognitive activities.

In addition to attentional problems, children with ADD usually have problems with impulse control in both the cognitive and behavioral areas. Two example of behavioral impulsivity are blurting out the answers to questions before the teacher has finished asking the questions, and butting into other children's games without first asking to be allowed to join in. An example of cognitive impulsivity would be doing a whole set of problems incorrectly based on a quick first impression (e.g., doing addition on a series of multiplication problems, based upon the impression that the "x" is a plus sign).

It is currently thought that there are three subtypes of ADD: *"attention deficit disorder with hyperactivity"* (ADDH), *"attention deficit disorder without hyperactivity"* (ADDW). This means that not all children who have ADD have the cluster of hyperactivity symptoms (which, as noted above consists of symptoms such as fidgeting, squirming, having difficulty remaining seated, excessive running around or climbing on things, moving about excessively during sleep, and always being "on the go"). Recent data suggests that children with ADDH are different from children with ADDW in other ways as well as with regard to the hyperactivity symptom cluster (Cantwell & Baker, 1992).

It was once estimated that between 3 and 5% of school-aged children had ADD. However, it now appears that the disorder is much more common. In a currently ongoing study being conducted at Yale University (Shaywitz et al., 1988), one out of every five youngsters in kindergarten through third grade classes were found to have ADD symptoms.

This study is also finding that, while ADD is more common in boys than in girls, there are many more girls with this problem than was once thought. For some reason, the girls are not referred for professional evaluation as often as boys, even though they are still there in the community.

The lower referral rate for girls may be because girls have fewer associated problems such as oppositional, negativistic, or aggressive behaviors. These kinds of behaviors, which are common in boys with ADD, are highly disruptive and therefore likely to result in referral for professional help. Girls who have ADD tend not to have these behaviors, and, instead, tend to be quitely inattentive and to have learning problems.

The Yale study also is finding (Shaywitz et al., 1990) that learning disabilities are equally common in girls as in boys, at least in young (kindergarten through third grade) children. However, learning disabled boys are more likely than learning disabled girls to be referred for professional evaluation. Again, this appears to be the result of associated problems (such as disruptive classroom behaviors) in the boys.

ADD is particularly likely to be present in children who have learning problems, speech/language disorders, and other forms of psychiatric problems (Baker & Cantwell, 1992; Huessy, 1992). The psychiatric problems that most often occur along with ADD are: conduct disorder (juvenile delinquency), oppositional disorder, and depression. In fact, the majority of children who have ADD also have other programs or diagnoses in addition to the ADD. Naturally, the presence of such other *"comorbid"* disorders affects the particular kind of intervention that is most appropriate for a child.

ADD is a condition that is probably present at conception. We often hear mothers of ADD children say things like *"Johnny was hyperactive in the womb"* or *"The day Johnny was born, I put on my Reeboks and I have been running ever since."* Parents are often able to identify problems in the areas of attention, poor impulse control, and an excess of motor activity, during the toddler years. The younger a child is, however, the more difficult it is to make an accurate diagnosis. This is because short attention span and poor impulse control are typical in young children.

As babies, children with ADD may have been colicky. They often have what Stella Chess (1979) has called a *"a difficult temperament,"* that is, they don't sleep very well, they don't maintain a day/night pattern very well, and they are very difficult to soothe. Children with ADD have high rates of allergies (around 25%), which may make the ADD behaviors worse, partially because allergy medications (e.g., steroids, inhalers and antihistamines) may cause more hyperactivity.

ADD symptoms are very persistent over time, and, contrary to what was once believed, ADD children do not outgrow the disorder at puberty. Often, in fact, ADD symptoms may become worse in adolescence. Although impulsivity and attention generally improve as the child gets older, the majority (approximately 80%) of ADD patients do not ever become completely normal in these areas. As a result, adolescents with ADD typically do not well in school, and, later, when they enter the job market, entry tends to be at a lower level than for their peers.

The most common outcomes of ADD, if not treated, is the development of additional problems, most often learning disabilities, antisocial behaviors, and substance abuse (usually alcohol abuse). One follow-up study (Satterfield et al., 1982), found that 25% of ADD teens had been incarcerated by the age of 18, and 55% had been arrested for committing a felony.

The question of whether ADD persists into adult life has only recently been examined. It appears that between 40% and 66% of cases of childhood ADD do persist into adult life.

The term *"ADD, residual state"* (ADD-RS) is used to describe adolescents and adults who had ADDH when they were younger. These individuals have attentional symptoms, but different attentional symptoms from those seen in ADD youngsters. The adults have an *"internal fidgetiness"* and can't relax. They overreact to environmental stimuli, tend to have trouble sitting still, and have histories of multiple job changes.

INTERVENTIONS FOR ADD

As stated above, treatment for ADD involves a *"package"* of interventions. Intervention must begin with a complete evaluation, including assessment of behaviors at home and at school, assessment of academic achievement performance, and assessment of related areas of development (e.g., speech, language, motor coordination). Once a child's deficit areas are identified, an appropriate treatment package can be created using the relevant intervention strategies such as educational assistance, medications, linguistic/cognitive interventions, parent management training, behavioral training, and social skills training.

School placement is probably the most important intervention consideration, because it affects the child five days a week, six hours a day. Although most ADD children go to regular schools, special considerations (such as small class size) need to be made. Because school placement is very crucial, it should be re-evaluated at the end of every school year to ascertain that there is always the best possible *"fit"* between the child and his classroom. Other kinds of educational intervention (such as one-to-one special tutoring) may also be needed.

Our research (Baker & Cantwell, 1992) indicates that as many as 50% of ADD children also have speech or language disorders. For these children, speech/language therapy is useful. Parent management training can be very helpful because with ADD children especially good parenting skills are needs.

For some ADD children, cognitive behavioral therapy can prove beneficial in those behavioral areas where medications might not be effective. Cognitive behavioral therapy trains the child to think differently, or to use different cognitive strategies (Braswell & Bloomquist, 1991).

Social skills training is useful for those ADD children who have a lack of social *"savoir faire"* and do not recognize the important nuances of social interaction. Social skills can be taught individually, in a small group setting or in a larger special classroom (Swanson et al., 1990).

MEDICATION INTERVENTIONS FOR ADD

In order to understand how psychiatric medications work, some understanding of neurophysiology and brain functioning is needed. Basically, electrical impulses are transmitted from the brain down the nerve cell. There are gaps between one nerve cell and the next one (*called "synapses"*) and certain chemicals in the brain *(called "neurotransmitters")* act as messengers from one brain cell to another across the synapses. There are a variety of these neurotransmitters, for example, Serotonin, Norepinephrine, and Dopamine. The psychotropic medications that alter ADD symptoms do so by acting on these various neurotransmitters.

All ADD children do not respond to medication. However, the evidence is overwhelming that medication with any other kind of therapeutic intervention is better than medication without any other interventions or than other interventions alone. Therefore, it is a disservice to the average child with ADD not to try medication.

Generally, medication is very effective for managing many of the symptoms of most ADD children. Different medications (discussed below) have different strengths and weaknesses. However, there are some issues that must be considered with all forms of medication.

One issue is that many children are negative about taking medication. This is especially true at adolescence, which is a time of body-image formation when there is a strong emotional need not to be *"different."* Taking a drug may confirm to youngsters that they are different, and consequently, many are non-compliant about taking medications. In fact, approximately 50% of adolescents are non-compliant about taking psychotropic medications.

We have found that it is sometimes helpful to explain the medication as a *"handicap deterrent."* A youngster can be told that, just as some people have to wear reading glasses in order to correct a visual deficit, so some people need medications in order to improve an attention or concentration deficit. The youngster can also be told that taking the medication is a short-term situation, like wearing a splint or a cast for a sprained arm or broken leg. However, most ADD children will require medication throughout adolescence (Clampit & Pirkle, 1983). As the child moves through adolescence into adult life, the dosage (on a milligram per kilogram of body weight basis) can often be decreased.

Another issue with psychotropic medications is that they need to be monitored carefully. If a child is on medication with a stable dose for a long time, monitoring can be done every six months or so by telephone. However, if the medication is being titrated, it needs to be monitored very closely (e.g., at least on a weekly basis).

Children themselves are notoriously bad at reporting medication effects. Even adolescents who have higher levels of cognitive functioning, and who are more reflective about themselves, generally cannot begin to describe what medication is doing. In fact, many of them deny that the has any effect, and may simply attribute medication effects to their own *"trying harder"* to concentrate.

Rating scales or parent and teacher reports are the best ways of monitoring drug effectiveness, since these give quantitative, objective indications of how the child performs in both the educational and home environments. Rating data is needed for *"baseline"* (i.e., before the medication treatment is started) as well as on various doses of medication.

The third issue with medication is that there may be major short-term changes in behavior when the ADD child is first given the medication. For example, the child may have an increased sensitivity to environmental noise (e.g., an air conditioning unit), and he becomes even more active for a few days or weeks. Possibly these short-term changes are due to the neural receptor sites becoming attuned to the medication.

There are two general types of medication that are very effective for the treatment of ADD: stimulants and tricyclic antidepressants. Other medications are also sometimes used for the treatment of ADD. These types of education are discussed separately below.

Stimulants

Stimulants are the most commonly used medication for the treatment of ADD. Currently, four major types of stimulants are used: methylphenidate (or Ritalin); amphetamines or (Dexedrine); magnesium pemoline (Or Cylert); and fenfluramine hydrochloride (or Pondimin), which is only used infrequently.

Approximately three-quarters of all ADD children respond positively to at least one of the stimulants. The positive effects of stimulants include: decreases in motor activity, improvements in sustained attention, decreases in impulsivity and distractibility, improved motivation, and improvements in accuracy and speed of academic achievement performance (Dulcan, 1990; Kavale, 1982).

Overall, the three major stimulants (i.e., Ritalin, Dexedrine and Cylert) have about the same level of effectiveness. However, many children will respond positively to one stimulant and not to another stimulant. The reason for these differential responses is that the different stimulants act upon different neurotransmitter systems. For example, amphetamines effects primarily norepinephrine, and methylphenidate primarily effects dopamine. Cylert effects both norepinephrine and dopamine, but less strongly than the other two drugs. Pondimin lowers bloods serotonin, and probably brain serotonin as well.

Although we know that individual respond differently to the various stimulants, we do not know to predict individual responses. Thus, we cannot predict whether a particular child will respond better to one medication than to another, or whether he will have fewer side effects with one drug than another. As a result, treatment must use a trial-and-error approach. First one medication must be tried at a low dosage, and then the dosage must be gradually titrated up while assessing whether there are positive clinical effects and/or negative side effects.

It was once thought that stimulants were responsible for growth suppression in children. However, prospective studies have shown that there are no long-term effects on either height or weight (Beck et al., 1975; Hechtman et al., 1984). In fact, stimulants have been used to the last 50 years with ADD children and are considered to be both effective and safe (Cantwell, 1980). Common side effects include increased heart rate, increased gastro-intestinal movements (stomachaches), and headaches. Rarely, the medication may cause tics. However, in clinical practice, it is extremely rare to have to discontinue stimulant treatment because of side effects. A failure to respond to the medication of a more common clinical problem.

One of the drawbacks of the stimulants is that they are short-acting. Even the so-called *"long-acting"* or *"slow-release"* forms last only six to eight hours. This means that, by the end of the day, ADD children who were given stimulants in the morning will no longer show a positive response. And, when these children wake up the next morning, it is as if they had have never had any medication; there is no build-up of the medication, and there is no carry-over of effects from the days before.

Sometimes clinicians recommend reducing the dosage of stimulants or taking the child off the medication entirely during the weekends or summer vacation. This practice, called a *"drug holiday,"* was once very popular, and was motivated by the belief that youngsters would take less medication over time and therefore would be less effected by growth suppression. However, research has shown that neither the stimulants nor the *"drug holiday"* have significant effects on long-term growth rates.

Consequently, we feel that the only good rationale for a drug holiday is that a child does not need the medication. Frequently, a child will need the medication on weekends and holidays. For example, an ADD child who is involved in weekend sports such as softball, will benefit from medication by its reduction of *"off task"* behaviors (e.g., watching the birds or kicking the grass) during games.

Furthermore, restarting the medication after a drug holiday can result in the child suffering a *"break in"* period, just as if he were starting the medication for the first time. Agitation, decreased appetite, and/or feelings of fatigue are short-term symptoms that may occur during this break in period. Presumably these symptoms are due to the neural receptor sites reacting to the drug as if it was a new drug.

Tricyclic Antidepressants

Tricycles are affective alternative medication treatment for ADD, although their primary use in this country is for treating adult depression. Standard tricycles (including imipramine [or Tofranil] and amitriptyline [or Elavil]) have been in use since 1957, and there are numerous other tricyclics available as well.

Some tricyclics primarily act upon norepinephrine; others primarily act upon serotonin; and still others primarily act upon dopamine. Because of their actions upon these neurotransmitters, the tricyclics can affect ADD symptoms in ways similar to the stimulants. Specifically, they can increase attention span, improve impulse control, decrease fidgetiness, and decrease restless behaviors. Some tricyclics (e.g., clomipramine) are also effective in reducing obsessive-compulsive behaviors, and therefore may be the best drug for those ADD children who have obsessive-compulsive problems.

Although the tricyclics share some effects in common with stimulants, they are different from the stimulants in three major ways. First, tricyclics are long-lasting and last all day. This means that once a certain level of tricyclic medication is built up in the blood stream, the effects remain constant. There isn't the *"up and down"* phenomenon that occurs from morning to evening with stimulants.

Second, tricyclics do not produce stimulant side effects such as appetite suppression, stomachaches or tics. Consequently, tricyclics may be preferred medications for those ADD children who have developed eating problems from stimulants, or who have a tic disorder which might be exacerbated by stimulants.

Third, tricyclics have both anti-depressant and anti-anxiety effects. Thus, they may be the preferred treatment for an ADD child who has those other symptoms as well as ADD symptoms.

For the average ADD child, however, stimulants are a better choice simply because more is known about them. Stimulants are also easier to monitor because they are short acting. For example, one can determine the effect of particular dose of stimulants within a few days. In contrast, it takes at least a week for the tricyclics to build up in the blood stream to the point where their effect can be measured. Also, because of the time needed for tricyclics to build up in the blood stream, drug holiday are not possible with tricyclics. For example, if tricyclics are not administered during the weekend, their level in the blood drops so that the following Monday, there will be no positive effect from taking the medication again.

Other Medications For ADD

Clonidine is another medication that may be useful for ADD. It was first used at Yale as a treatment for Tourette's Syndrome (TS), a tic disorder (Cohen et al., 1992; Leckman et al., 1991). For those TS patients who also had ADD, in improvement in the ADD symptoms was observed with Clonidine.

A very small number of ADD patients who do not have TS have been treated with Clonidine (Hunt et al., 1984). Most of these patients showed a positive response to the medication, but the nature of the response was somewhat different than that which occurs with stimulants or with tricyclics. Also, Clonidine is less effective than stimulants for the cognitive manifestations of ADD. However, for those ADD children who are also aggressive-disruptive, or who also have tics, it may be a good medication choice.

One problem with Clonidine, however, is that it is extremely short-acting, and in many cases has to be given as many as five times a day. There is a long-lasting patch available, but children like to pull it off.

Bupropion (or Wellbutrin) is a new antidepressant that may be somewhat effective for certain children with ADD. Prozac is another new antidepressant that may also be effective for certain individuals with ADD. Although little is known about its effects upon children, when used with adult ADD-residual state patient, it often has a dramatic positive effect.

MOOD DISORDERS

"Depression" and *"demoralization"* are two mood problems that are very common in learning disorder adolescents. *"Depression"* can occur in various forms (e.g., major depression and dysthymia), sometimes alternating with periods of *"elation"* or *"mania"* (bipolar disorder, cyclothymic disorder). One striking feature of depression is *"anhedonia,"* or an inability to get pleasure from the things that ordinarily give pleasure.

Mood disorders are extremely common in the general population, particularly in females. One woman in four will have a significant episode of depression in her lifetime, as will about one man in ten. This gender differences does not occur until after puberty. Prior to puberty, the rates of depressive disorders are exactly the same for boys as for girls. After puberty, the rate is twice as high in girls as in boys.

Depressive disorder are a different from ADD \in that ADD is a chronic problem that is always there, whereas depression is an episodic problem that can be resolved. Medications do not cure ADD; they simply alleviate many of the symptoms. If the medications are discontinued, the disorder will still be present. For depressive disorders medication can be withdrawn after treatment, and the condition may no longer be present. However, relapse may occur with time, necessitating retreatment.

A number of medications are effective for depressive disorders in children and adolescents. The tricyclic antidepressants (discussed above) and the newer non-tricyclic antidepressants (e.g., Prozac and Wellbutrin) may help. When tricyclics are used for depression, different dosages are given than when the drug is used to treat ADD.

For manic symptoms, Lithium and Tegretol are two medications that may be useful. Lithium, as an anti-aggressive in those children who exhibit isolated outbursts of physically aggressive behaviors associated with significant mood swings. However, children who are *persistently* aggressive, do not show this response to Lithium.

The side effects of Lithium include fine hand tremors, and gastrointestinal problems (i.e., nausea, diarrhea, vomiting). These side effects are related to the concentration of the drug in the blood stream and indicate when the dosage is too high.

"Demoralization" (or bad feelings resulting from failure in a important area of life) is a common mood problem, especially in children with learning problems. *"Demoralization"* is not *"depression;"* it is not characterized by *anhedonia,"* and it does not respond to antidepressant (or other) medications.

The treatment for demoralization is double-pronged. First, the underlying problem that caused the demoralization must be remedied. For example, if the child is demoralized by a severe learning problem and academic failures, then he needs to be helped to succeed academically. In the meantime, cognitive-behavioral interventions are helpful. This form of therapy, first used many years ago for depression (Beck, 1967) works by changing cognition, reversing negative views of the self, of the world, and of the future.

ANXIETY DISORDER

A final category of psychiatric disorders that are common in children and adolescents, (particularly those with learning disorders) is *"anxiety disorders."* Anxiety disorders exist in several forms including: *"separation anxiety disorder," "avoidant disorder," "generalized anxiety disorder," "post-traumatic stress disorder" (PTSD),* and *"obsessive-compulsive disorder"* (OCD).

"Separation Anxiety Disorder" consists of an abnormal fear of being separated from mother, the home, or a familiar environment. Although separation fears are considered normal in toddlers, they are not normal in school aged children. *"Separation Anxiety Disorder"* can be manifested by a refusal to go to school, a refusal to sleep over at somebody's house, by not letting the parents leave the house, or by requiring parents to sleep with the child. The term *"school phobia"* has been used for the disorder, but that is not appropriate since these children do not have a true fear of school.

Medication alone is usually not an effective treatment for *"Separation Anxiety Disorder"*. A behavioral intervention program is needed in order to get these children to separate comfortably. However, anti-anxiety drugs (i.e., Imipramine, or Xanax) can be beneficial in combination with the behavioral program.

"Avoidant Disorder" is a disorder in which, (despite some close peer relationships), there is excessive shyness, fear of social situations, and fear of making friends. This condition definitely does not respond to medication.

"Generalized Anxiety Disorder" is characterized by a wide range of symptoms reflecting fears in a variety of different kinds of settings. This condition is also not usually responsive to medication.

"Post-Traumatic Stress Disorder" is a reaction to a severe trauma, such as witnessing a murder or gang violence. Medications such as Clonidine

and Propranolol and some antidepressants can decrease anxiety in children with this disorder. However, these are adjunctive treatments. The most effective intervention for this disorder is desensitization involving a reenactment and rebuilding of the traumatic event.

"*Obsessive-compulsive Disorder*" (OCD) is an anxiety disorder characterized by obsessional ideas or images and ritualistic, compulsive, and repetitive behaviors. Typical symptoms include lining up objects or counting objects; having excessive fears of germs or illnesses; and engaging in excessive hand washing. the disorder is very difficult to diagnose in children, because they will recognize that their symptoms are "*crazy*" and will be very secretive about them.

As recently as five years ago, it was believed that OCD was unresponsive to medications. However, the advent of the serotonin reuptake blockers (i.e., Anafranil and Prozac) has dramatically altered the treatment of OCD. These drugs do not cure OCD, but they make the condition substantially better. In combination with the psychotherapy, these medications can help these children to live a relatively normal life. Anafranil (or clomipramine) is particularly effective for OCD, having dramatic, striking effects on obsessive-compulsive behavior in approximately 80% of children. It is dramatically different from its parent compound (Imipramine) which has no effects on obsessive-compulsive symptoms.

REFERENCES

Baker, L., & Cantwell, D. P. (1992). Attention deficit disorder and speech/language disorders. *Comprehensive Mental Health Care, 2*(1), 3–16.

Beck, A. T. (1967). *Depression: Clinical, Experimental & Theoretical Aspects.* New York: Harper & Row.

Beck, L., Langford, W. S., Mackay, M., & Sum, G. (1975). Childhood chemotherapy and later drug abuse and growth curve: A follow-up study of 30 adolescents. *American Journal of Psychiatry, 132,* 436–438.

Bell, L., & Bloomquist, M. L. (1991). *Cognitive-behavioral therapy with ADHD Children: child, family, & school interventions.* New York: Guilford Press.

Cantwell, D. P. (1980). A clinician's guide to the use of stimulant medication for the psychiatric disorders of children. *Journal of Developmental & Behavioral Pediatrics, 1*(3), 133–140.

Cantwell, D. P. & Baker, L. (1992). Attention deficit disorder with and without hyperactivity: a review and comparison of matched groups. *Journal of the American Academy of Child and Adolescent Psychiatry, 31,* 432–438.

Chess, S. (1979). Development theory revisited: Findings of a longitudinal study. *Canadian Journal of Psychiatry, 24,* 101–112.

Clampit, M. K. & Pirkle, J. B. (1983). Stimulant medication and the hyperactive adolescent: Myths and facts. *Adolescence, 28,* 811–822.

Cohen, D. J., Riddle, M. A., & Leckman, J. F. (1992). Pharmacotherapy of Tourette's syndrome and associated disorders. *Psychiatric Clinics of North America, 15*(1), 109–129.

Dulcan, M. K. (1990). Using psychostimulants to treat behavioral disorders of children and adolescents. *Journal of Child and Adolescent Psychopharmacology, 1*, 7–20.

Hechtman, L., Weiss, G., & Perlman, T. (1984). Young adult outcome of hyperactive children who received long-term stimulant treatment. *Journal of the American Academy of Child Psychiatry, 23*, 26–1269.

Huessy, H. R. (1992). Comorbidity of attention deficit hyperactivity disorder and other disorders (letter). *American Journal of Psychiatry, 149*(1), 148–149.

Hunt, R. D., Cohen, D. J., Anderson, G., & Clark, L. (1984). Possible change in noradrenergic receptor sensitivity following methylphenidate treatment: Growth hormone and MHPG response to clonidine challenge in children with attention deficit disorder and hyperactivity. *Life Sciences, 35*(8), 885–897.

Kavale, K. (1982). The efficacy of stimulant drug treatment for hyperactivity: A meta-analysis. *Journal of Learning Disabilities, 15*, 280–288.

Satterfield, J. H., Hope, C. M., & Schell, A. M. (1982). A prospective study of delinquency in 110 adolescent boys with attention deficit disorder and 88 normal adolescent boys. *American Journal of Psychiatry, 139*, 795–798.

Shaywitz, S. E., Shaywitz, B. A., Fletcher, J. H., & Escobar, M. D. (1990). Prevalence of reading disability in boys and girls: Results of the Connecticut longitudinal study. *Journal of the American Medical Association, 264*, 998–1002.

Shaywitz, S. E., Shaywitz, B. A., Schnell, C., & Towle, V. R. (1988). Concurrent and predictive validity of the Yale Children's Inventory: an instrument to assess children with attentional deficits and learning disabilities. *Pediatrics, 81*(4), 562–571.

Swanson, J., Simpson, S., Agler, D., Kotkin, R., Pfiffner, L., Bender, M., Rosenau, C., Mayfield, K., Ferrari, L., Holcombe, L., Prince, D., Mordkin, M., Elliot, J., Minura, S., Shea, C., Bonforte, S., Youpa, D., Phillips, L., Nash, L., McBurnett, K., Lerner, M., Robinson, T., Levin, M., Baren, M., & Cantwell, D. (1990). UCI-OCDE school-based treatment program for children with ADHD/ODD. In C. N. Stefanis, A. D. Rabavilas, & C. R. Soldatos (Eds.), *Psychiatry: A world perspective*, (vol. 1, pp. 1107–1012). Elsevier Science Publishers B. V.

LECTURE 9

Joan T. Esposito
Dyslexia Awareness and Resource Center, Executive Director
California Learning Disabilities Association, Past President

THE EMOTIONAL AND EDUCATIONAL CHALLENGES OF DYSLEXIA AND ATTENTION DEFICIT DISORDER: ONE STORY

"I feel trapped and alone in a shell of dark sadness, surrounded by laughter. I always feel alone—trapped in a room of darkness."

—Ten-year old boy
with learning disabilities

Introduction by Richard L. Goldman

For our last seminar, our speaker will address the challenges of having dyslexia and attention deficit disorder.

As an adult with attention deficit disorder and dyslexia, and whose son also has dyslexia, Ms. Esposito brings a unique perspective. She could barely read until she was 44 years old and has experienced all the frustrations and challenges of dealing with a learning disability.

She was president of the California Learning Disabilities Association for 2 years and previously served as their Governmental Affairs Chairperson. She also has received numerous national awards, including a nomination as one of the President's Thousand Points of Light. Ms. Esposito is the Founder and Executive Director of The Dyslexia Awareness and Resource Center in Santa Barbara. The center provides information and support services to dyslexics of all ages. Most importantly, she has turned her personal experience into a crusade by being a committed advocate at the state and national level.

It is my pleasure to introduce Ms. Joan Esposito.

Ms. Joan Esposito

Because I have both dyslexia and attention deficit disorder, I need to read my presentation to you. Because of my learning disabilities, I have trouble retrieving information quickly from my memory, and I have great difficulty organizing and expressing my thoughts in sequence. I may start at the end of a story, jump to the beginning, and end in the middle. Often I leave out important details that I feel I have already shared, but, somehow, the words have remained in my mind.

Eight years ago, my reading proficiency was at a fourth grade level. My spelling was atrocious; I had first grade punctuation and grammar skills. I was not identified as dyslexic until age forty-four after my seventeen-year-old son was diagnosed with dyslexia and attention deficit disorder. I grew up in Liverpool, England after World War Two. I was the fifth of six children. My father had dyslexia, while my mother showed no signs of learning disabilities. Although undiagnosed, several members of my family have symptoms of learning disabilities.

"When you walk through a storm hold your head up high and don't be afraid of the dark." I sang these words from a popular song over and over to myself when I was a teenager. I lived in a storm all of my life until I learned at age forty four that my reading and spelling problems were the result a condition called dyslexia. As a young child in England, the time that I spent attending classes and attempting to learn was literally hell. Every morning, I woke up sick to my stomach at the thought of school. I could not understand why my parents made me go to school every day and struggle. I simply could not learn—no matter how hard I tried. My teachers could not teach me.

Because I did not learn like many of my classmates, I did not socialize with them either. How could I? I could not read or spell like them. I was constantly teased by them. I did not play with the others because I felt different from them. I could not understand or explain why I felt different, I just did! Even without a name for the difference, I knew deep inside that I was different. Each day I sat in class and prayed that the teacher would not ask me to read out loud. I went home from school at night and cried myself to sleep because I did not understand why I could not read or spell as well as my classmates. As I tried to spell and write legibly I told myself:

"I don't look retarded but I must be slightly retarded."

"Maybe I have brain damage."

"The teachers say I can spell if I try harder, but I do try and it does not work."

"Maybe I can't spell because I was born during the war while they were bombing Liverpool and somehow the noise of the bombs affected the way I can learn."

"Maybe I can't spell because I am the fifth of six children and they got all of the brains from my parents and left none for me."

"I must never have any children in case they inherit my damaged brains."

"Maybe I will die before I leave school and then I will not have to spell."

I spent hours alone in my room, trying to figure out how to hide my reading and writing problems from my family and friends. I lied and cheated my way through school.

At the age of fifteen I went to work in a cigarette factory, brushing floors. At seventeen I left home and moved to London, where I worked as a chambermaid in a West End hotel, until a friend gave me a job in his office. Throughout my life my jobs all were given to me by close friends. I was able thus to avoid filling out job applications. I came to America as a nanny in 1963, when I was twenty years old. With the help of a friend who read the textbook to me, I later became a manicurist. In 1968 I married a literary agent from Beverly Hills. Our marriage lasted eleven years. During that time I entertained some of the top studio executives, directors, producers and actors in Hollywood.

The first week I was married, my husband told me that we were going to have some of his clients to our home for dinner: German actress Elke Sommers, her husband, Joe Hyams, and Elliot Silverstein, who had just successfully directed "A Man Called Horse" with Michael Cain. I confessed to my new husband that I was not a good cook, but he directed me to several cookbooks in the kitchen. What I didn't confess was that I couldn't read them. Somehow I managed to talk my new husband into cooking the meal. I helped and watched what he did, and soon learned how to cook what he could cook. To keep my reading problem a secret, however, I took a French cookery class. I watched the chef prepare a meal, then I went home and immediately cooked the same meal to remember it. To reinforce my memory I cooked and served the same meal over and over. Because of all the butter and cream sauces in French cooking, my husband developed a problem with gout. I now can smile at my contribution to his problems with gout you, and will see why as my story continues.

During the first part of my marriage I took dozens of tennis lessons. I had a strong serve, my coordination was good, and I was able to hit the ball where I wanted it. But, I could never remember the score or where I was supposed to stand. Elke Sommers and I had became good friends, and she was also taking tennis lessons. One day, as I was watching her take a lesson, she asked me to play a game with her. I quickly replied that I had given up my lessons and that she needed some one with more experience. I then had to stop taking lessons so that she would not discover my lie.

One success for me has been bargain hunting for antiques. Over the years, people have given me over forty books on antiques, but, of course, they were of no use to me. I learned by going to antique auctions. I went to a preview the day before each auction, where I would touch and feel the pieces and ask the auction attendant questions. I would then attend the auction. By watching the buyers in the audience bid and by writing down the prices at which the antiques went, I was able to remember both the antiques and their prices. I then could go to the antique shops and compare the prices.

Because of my problems due to undiagnosed dyslexia, I missed many business opportunities over the years. For example, Elke admired the antiques I bought, and asked me to go to Europe to buy antiques with her money and split the profits with her. I told her I could not leave my young son. What I could not tell her was that I could not read or write well enough to fill out the forms to get antiques back into this country. As a result of my learning disabilities I often get disoriented when I travel to new towns; even walking through new airports and catching a plane can be confusing for me. Not only the antiques would have been lost in Europe: I would too. Liza Minelli also admired my decorating talents. She tried to talk me into decorating her home. But once again I had to lie. I said I was too busy entertaining my husband's clients and looking after our young son.

One afternoon, I was sitting in our den with Liza, when my five-year-old son came in with a container of popcorn in a tinfoil dish. He wanted me to cook it on the stove top. I knew I could not follow the directions on top of the popcorn container. My husband, who usually made the popcorn, was in the other room talking business with Liza's future husband, Jack Haley, Jr. So I could not disturb him. I made several excuses to my young son, but he was not going to let me off that easily. When Liza saw that I was not responding to Joel, she picked him up and carried him and the popcorn into the kitchen. I did not dare follow them, in case she gave the popcorn back to me. So I sat in the den feeling stupid and sick to my stomach. Soon I could smell something burning, so I ran into the kitchen. Joel was sitting on the kitchen counter, and Liza was singing and dancing around the kitchen for him, while the popcorn burned on the stove.

On two occasions I was a guest in Charlton Heston's home, and I met him several times at the studios and at social functions. I found him a warm and wonderful person but tried my best to avoid talking with him because of his dry sense of humor. As a result of my learning disabilities, I don't always understand jokes or anyone with a dry sense of humor. Telling jokes was Chuck's way of making me feel comfortable whenever we met, but his humor, ironically, had the opposite effect on me.

Before my dyslexia was diagnosed and I understood how I function as a person with dyslexia, I always felt uncomfortable at parties and in other social situations. I would avoid parties like the plague. But on one occasion, I allowed myself to be talked into going to a Tupperware party at a friend's house. Towards the end of the party we were to play a game for which we had to write down five nouns. I quickly excused myself and went to the bathroom. I walked back and forth, looking in the mirror, and told myself how stupid I was. I had been told over and over what a noun was; why could I not remember? I felt sick to my stomach. I stayed in the bathroom as long as I could and hoped that they had finished their game without me. I came out of the bathroom to find they had waited for me, so that I would not miss the chance to win the prize, a plastic cup. I can't remember how I escaped, but I can assure you I never went to another Tupperware party.

I worked hard and became very clever at covering up my problems with language, but it took a toll on my health. I developed severe hypoglycemia. Several times, I was admitted to the hospital for violent stomach pains and headaches. Each time, after numerous tests showed no physical cause for the pains, I was released from the hospital without a diagnosis.

After several years of struggling to entertain clients, I was thrilled at the idea that our family would leave Beverly Hills for Santa Barbara. We had no clients or friends in Santa Barbara, and there I could hide away from the world. We bought a large, old, Spanish home. For the first six months in Santa Barbara I was happy. I spent the time doing things I loved to do, things I was gifted in: gardening, remodeling and decorating our home. Then, one day, my son's teacher asked me to help in his second grade classroom. I thought I would be able to work with the children on their art projects or just watch over them for the teacher. But the teacher asked me to help the children with their spelling and reading. I made an excuse to leave the classroom, and I never went back.

At this point in life I became a recluse. I would not answer the telephone. I very seldom left the house. When friends came from Beverly Hills to visit, I would stay in my bedroom and pretend that I was ill. I saw very few of our old friends. More and more, I withdrew into myself and became deeply depressed. I knew that something was wrong with me, but did not know where to go to get help. I went to a doctor and a counselor, and they both blamed my withdrawal and depression on my lack of an education, my low self esteem and my domineering husband. The doctor's solution was to put me on a heavy dose of anti-depressants, which made me sleep most of the day.

My ex-husband, my son's father, also has severe symptoms of dyslexia and attention deficit disorder, although he has not been formally diagnosed. He brought the anger, frustration and pain that he had experienced in school into our marriage, and I was the recipient of his violent temper and mood swings. I was not unlike other women, unable to acquire an education because of their dyslexia, afraid to leave an abusive husband for fear of not being able to get a job to support their children. After eleven years of marriage I could no longer take his physical abuse, and it was starting to affect our son. Somehow, I worked up the courage to file for a divorce.

During divorce proceedings, our home sold for over one million dollars, and we had another million and a half in assets. From the day my nine-year-old son and I moved out of our home, we were homeless off and on for six years. Fourteen months from that day we found ourselves on food stamps. My husband had handled all of the finances and bank accounts.

My husband and his lawyers used my lack of education against me in court to try to gain custody of our son. The probation officer in our child custody case reported to the court that he found that "although Joan is uneducated, I find her quite intelligent." Some of the most humiliating experiences of my life came when I sat in depositions and courtrooms full of strangers while my illiteracy and lack of a formal education was brought up over and over again in reference to my gaining custody of our son. Testimony on my illiteracy was used over a seven year period, in over two hundred court appearances.

One year into the divorce proceedings my husband filed bankruptcy and he put my assets into his bankruptcy. He left the country with the rest of our assets and left me to pay the taxes, his creditors and his legal expenses from before and after our separation. He moved to England, where he met his next wife. She was a former Russian ballerina with the Royal School of Ballet in London. He bought a Manor House that had belonged to the late Lord Butler. He drove a Rolls-Royce and traveled around the world with members of Lord Sainsbury's family. He sent postcards to our ten-year-old about his travels abroad and his hunting weekends at Lord Sainsbury's country estate. Meanwhile, I was living in Santa Barbara. I did not have the necessary skills to get a job with a livable wage, so I cleaned hotel rooms to support us. The salary I made was not enough to even pay our rent, but we were able to survive with the help of friends. (You can see now why my ex-husband deserved gout.)

At one point in the bankruptcy court proceedings, when I had no lawyer, the judge said I had to write a letter. I jumped up and said, "Your Honor, I don't know what it is, but I have problems writing letters. I need a dictionary to write!" He was not at all happy with my interruption and ordered me to write the letter. What the judge heard me say and what I thought I had said were totally different. I thought I had said, "Your Honor, I can't spell. My hand writing is unreadable. I do not know anything about grammar. It is very difficult for me to get my thoughts down on paper in sequence, and I can't find the correct spellings of words in a dictionary. I own seventeen dictionaries and can't find one that works for me." I truly believed that I had explained my problem clearly to the judge, because the thoughts were in my head. But they simply did not come out the way I thought they did. It was only when I read the court transcript, years after the divorce was final, that I realized I had not fully explained my language problem to the judge.

Besides my lawyers and my ex-husband, only two friends knew about my spelling and writing problems. I spent most of my days and evenings writing letters to my lawyers by hand. I rewrote my letters over and over; I made mistakes copying from one page to another. My hand and brain would get tired. The physical writing on each line and page looked different: I used print, script, upper and lower case letters, all in one sentence. I wrote descriptions of myself in the first, second and third person, as I still do. Writing was tiring and time consuming. I was constantly on the telephone to my two friends, asking them to spell words for me. After months of writing by hand I bought a typewriter for ten dollars at the swap meet. I was able to write longer letters, but they still had commas and periods wherever I felt like putting them, and paragraphs were nonexistent.

Because of the custody fight over our son and my fear of losing him, I gathered up courage and drove to the Santa Barbara City College to enroll in an English class. I was sent into a small building where I was given a test. The questions on the test directed me to find such things in a sentence as the object, the indirect object, the clause, or the prepositional phrase. I could not understand what an object had to do with grammar. An object to me was a thing. At that point I could not even pronounce the word "prepositional phrase." Santa Claus was the only meaning I could get from the word "clause," and "prepositional" was beyond my vocabulary. I slipped out of the testing lab when no one was looking and never returned to finish.

A few months after I married my new husband Les, we discovered that both my son and I had dyslexia and attention deficit disorder. Les helped me enroll for classes in the learning disabilities department at Santa Barbara City College. I was forty-four years old, functionally illiterate and full to the brim with low self esteem. At college I discovered that I *could* learn how to read and write and met other people who function like I do. Like me, my classmates had lived with a lifetime of pain. Some of the men would bang their fists on their desks out of frustration if they could not do their work. Others would get angry if another student laughed at their mistakes. They were not past the pain they had experienced in grade school when their peers or teachers teased them. Students expressed anger over their wasted years in grade school and high school. We questioned our teachers at the college: why could they teach us the reading, spelling, writing, and math that we could not learn in grade school? You could hear the anger in some students voices when they questioned the teachers: "WHY DIDN'T WE HAVE TEACHERS LIKE YOU IN OUR SCHOOLS WHEN WE WERE KIDS?!"

It was because of my own experiences and those of other students in my college class that I started the volunteer work I have been doing ever since. I approached our Santa Barbara newspaper with an article on dyslexia. After the article appeared in the paper, I was overwhelmed with calls from parents asking me to help them advocate for their children with dyslexia. After three years of working out of our home, in 1990, my husband Les and I founded the Dyslexia Awareness and Resource Center, which is a non-profit organization in Santa Barbara. We have assisted over six thousand clients, and all the services at the Center are free. I attend Individual Educational Plan (IEP) meetings with parents, expulsion hearings, court hearings and probation hearings for juvenile delinquents with learning disabilities. I meet with employers and teach them about dyslexia and how it affects employee performance. We train groups of counselors on the nature of learning disabilities. We have a seventy-thousand dollar library of books, teacher training tapes, dyslexia reading program videos, and video and audio tapes on learning disabilities, dyslexia, attention deficit disorder, and Tourette's Syndrome. The Center is the only one of its kind in the nation.

I can't finish my story without telling you something about my son, Joel, of whom I am so very proud. When he was seventeen, we discovered that he had dyslexia and attention deficit disorder. After years of struggling, he managed to graduate from high school. His report cards commented that

- "he needs to pay attention to written work, keep papers neat and show the depth of thinking that he indicates verbally";
- "sloppy and disorganized";
- "the weakness of his handwriting will probably preclude his joining the honors section of U.S. History next year"; and
- "your habits have shown every sign of being quite lazy; the result is that you are full of intriguing thoughts, insight and a fine vocabulary which you can only express in poor spelling, poor punctuation and sentence fragments or run-ons."

After high school my son entered the University of California in Santa Barbara with the most valuable tool in a dyslexic's life, a computer. He became a reporter and assistant editor for the university newspaper. He also wrote articles for local newspapers.

Joel has just turned twenty-three and is presently a freelance reporter in Yugoslavia, where he has been living for the last two and a half years. His news articles on Sarajevo are published in *Newsweek, The London Times, The Irish Times, The Washington Post, The San Francisco Chronicle, The Toronto Sun, The Miami Herald, The San Diego Chronicle* and numerous other national and international newspapers. His articles have been published in *Life, People* and *Rolling Stone* magazines, as well as several European magazines. He also reports live from Sarajevo for CNN television and Sky Television News in Europe.

I would like to close today with a quote from a man whom I admire for his work on behalf of learning disabled juvenile delinquents. Judge Jeffrey H. Gallet is a judge in the Family Court in New York state. Judge Gallet also struggled in school with dyslexia. He said, "IF YOU CANNOT READ THERE ARE ONLY TWO WAYS TO MAKE A LIVING—THE WELFARE SYSTEM OR CRIME—AND CRIME HAS MORE STATUS."

I would like to thank you for listening.

POSTSCRIPT

Before I finish, I just want to tell you how important it was for me to discover that I had learning disabilities. The labels I gave to myself as an uninformed and innocent child were debilitating. The diagnosis of dyslexia freed me to fulfill my dreams and become a functioning adult. I needed the appropriate label in order to find a teaching method that had worked for other people with dyslexia. Although I have several learning disabilities, including attention deficit disorder, the one that affected me most severely was dyslexia. I could read words that I had learned through whole word recognition, but with new words I struggled to match the sound to the written symbol on the page. *If you can't read, how do you learn about your other learning disabilities?*

Note: This chapter is part of a larger work in progress.

RESOURCES

ORGANIZATIONS

Association of Educational Therapists (AET)
14852 Ventura Blvd., Suite 207, Sherman Oaks, CA 91403
(818) 788-3850 Fax (818) 380-6896

Children and Adults with Attention Deficit Disorder (C.H.A.D.D.)
499 Northwest 70th Street, Plantation, FL 33317
(305) 587-3700

Council for Exceptional Children (CEC)
1920 Association Drive, Reston, VA 22091
(703) 620-3660

Council for Learning Disabilities
P.O. Box 40303, Overland Park, KS 66204
(913) 492-8755

Dyslexia Awareness and Resource Center
928 Carpinteria Street, Suite 2
Santa Barbara, CA 93103
(805) 963-7339

Etta Israel Center
6505 Wilshire Blve, Suite 503, Los Angeles, CA 90048
(213) 852-3222

Learning Disabilities Association (LDA)
4156 Library Road, Pittsburgh, PA 15234
(412) 341-8077

The Learning Disabilities Network
72 Sharp St., Suite A-2, Hingham, MA 02043
(617) 340-5605

National Center for Learning Disabilities (NCLD)
381 Park Avenue South, Suite 1420, New York, NY 10016
(212) 545-7510

Parents Educational Resource Center
1660 South Amphlett Blvd., Suite 200
San Mateo, CA 94402
(415) 655-2410 Fax (415) 655-2411

Orton Dyslexia Society
Chester Building, Suite 382, 8600 La Salle Rd, Baltimore, MD 21204
(410) 296-0232

The Rebus Institute
1499 Bayshore Blvd., Suite 146, Burlingame CA 94010
(415) 697-7424

PUBLICATIONS

ADHD Report, A bimonthly newsletter for clinicians edited by Dr. Barkley with contributions from leading clinicians and researchers. Call Guilford Publications at (800) 365-7066 to subscribe.

Answers to Distractions, by Edward M. Hallowell, M.D., and John J. Ratey, M.D., Pantheon Books, New York, 1994.

Attention Deficit Hyperactivity Disorder: A Handbook for Diagnosis and Treatment, by R. A. Barkley, Guilford Press, 72 Spring St., New York, NY, 1990.

Dr. Larry Silver's Advice to Parents on Attention Deficit Hyperactivity Disorder, by Larry Silver, M.D., American Psychiatric Press, Inc., Washington, DC, 1993.

Driven to Distraction, by Edward M. Hallowell, M.D., and John J. Ratey, M.D., Pantheon Books, New York, 1994.

I Can Learn, A handbook for parents, teachers, students. California Department of Education, Special Education Division, Sacramento, 1994.

Keeping A Head in School, by Dr. Mel Levine, Cambridge, MA, Educators Publishing Service, 1990.

The Learning Disabled Child: Ways That Parents Can Help, by Suzanne H. Stevens, N. Carolina, John F. Blair, 1985.

Learning Disabilities – A Family Affair, by Betty B. Osman, New York, Warner Books, 1979.

The Misunderstood Child, by Dr. Larry B. Silver, New York, McGraw-Hill, 1984.

No Easy Answers The LD Child at Home and School, by Sally L. Smith, Toronto, Bantam Books, 1980.

No One to Play With, by Betty B. Osman, New York, Warner Brooks, 1982.

Reading, Writing, and Rage, by Dorothy Fink Ungerleider, Rolling Hills Estates, CA, Jalmar Press, 1985.

Smart Kids With School Problems, by Priscilla L. Vail, New York, e.P. Dutton, 1987.

Succeeding Against the Odds, How the Learning Disabled Can Realize their Promise, by Sally L. Smith, New York, Jeremy Tarcher, Perigree Books, 1991.

Taking Charge of ADHD: The Complete Authoritative Guide for Parents, R. A. Barkley, Guilford Press, New York, 1995.

VIDEOTAPES

ADHD in Adults
ADHD in the Classroom
ADHD: What Do We Know?
ADHD: What Can We Do?
Russell A. Barkley, Ph.D.
Stonebridge Seminard, 1992, (508) 836-5570
Guilford Publications, 72 Spring St., New York, NY 10012
(800) 365-7006

How Difficult Can This Be?
Understanding Learning Disabilities (F.A.T. City)
Richard LaVoie, PBS Video, 1989. 800 344-3337, Fax (703) 739-5269

I'm Not Stupid
Gannett Broadcasting, Learning Disabilities Association
4156 Library Road, Pittsburgh, PA 15234 (412) 341-1515

Learning Disabilities and Social Skills:
Last One Picked, First One Picked On
Richard Lavoie, PBS Videos, 1994. (800) 344-3337, Fax (703) 739-5269

We Can Learn:
Understanding and Helping Children with Learning Disabilities
National Center for Learning Disabilities (NCLD), 1989.
(212) 545-7510, Fax (212) 545-9665

INDEX

T

U

V

W

9 783718 606238

For Product Safety Concerns and Information please contact our EU representative GPSR@taylorandfrancis.com
Taylor & Francis Verlag GmbH, Kaufingerstraße 24, 80331 München, Germany

www.ingramcontent.com/pod-product-compliance
Ingram Content Group UK Ltd.
Pitfield, Milton Keynes, MK11 3LW, UK
UKHW022203190926
475702UK00003B/183

* 9 7 8 1 1 3 8 1 8 1 2 2 9 *